D1805769

English for the
ENERGY INDUSTRIES
Oil, Gas and Petrochemicals

TEACHER'S BOOK

FIONA McGARRY

LANGUAGE
SOLUTIONS

Garnet
EDUCATION

Published by
Garnet Publishing Ltd.
8 Southern Court
South Street
Reading RG1 4QS, UK

www.garneteducation.com

This edition first published 2007

ISBN 978 1 85964 912 1

Production

Project manager: Francesca Pinagli
Editorial team: Francesca Pinagli, Maggie MacIntyre, Richard Peacock, Emily Clarke
Design: Robert Jones
Typesetting: Tina Shorter
Illustration: Doug Nash
Photography: Language Solutions

BP's Golden Rules of Safety on page 131 of the Course Book are used with the permission of BP plc.

Every effort has been made to trace the copyright holders and we apologize in advance for any unintentional omissions. We will be happy to insert the appropriate acknowledgements in any subsequent editions.

Audio production: Matinée Sound & Vision Ltd.

Printed and bound
in Lebanon by International Press

CONTENTS

INTRODUCTION

Who is the course for?

English for the Energy Industries is a course suitable for technicians, engineers and others working in the energy industries. It is aimed at people who work, or who are planning to work, on rigs and oil and gas fields, as well as those in workshops, laboratories, offices and transportation sectors of the petrochemical industry.

The course assumes that students will have an elementary level of English at the start of the course. It aims to provide practical help with technical contexts and vocabulary, as well as develop the listening, speaking, reading and writing skills necessary for coping in an English-speaking environment. It is also suitable for learners who have studied general English at school, but have either forgotten much of what they have learned, or do not feel confident in their ability to use English to interact with co-workers and supervisors. Such learners usually need a more technically specialized and practical course such as *English for the Energy Industries* to prepare them for using English in the workplace.

The course is designed to provide approximately 100–140 hours of teaching. It can be followed from start to finish as a course that builds knowledge step-by-step and revises and recycles vocabulary, structures and topics at the end of each section. Alternatively, teachers and students may wish to dip into it and focus on particular units and/or skills areas.

Approach and features of the book

The book focuses on language needed for practical tasks and communication in the workplace. Each unit title specifies a function that is given particular emphasis, starting with basic social and survival language in Unit 1. Later units cover more specialized key areas such as health and safety and the description of complex technical systems. There are plenty of opportunities to learn and practise useful language throughout the unit in communicative tasks such as information exchange and problem-solving activities.

Vocabulary is of paramount importance when using English in a technical environment. This book includes many specialized technical vocabulary terms and expressions, including abbreviations and colloquial expressions specific to the energy industries. Semi-technical terms are also given clear focus, for example the difference between controls such as dials, buttons, levers and switches is highlighted, along with their associated verbs. Clear contexts are used to introduce and practise new lexis. There are also motivating, but progressively more challenging, tasks that recycle and build on vocabulary, such as the wordsquares and quizzes in the review sections. Learners are also encouraged to categorize vocabulary, test one another on new words and practise looking up terms in the glossary at the back of the course book. This provides simple definitions of high-frequency technical words and phrases that make up key terms in the energy industries, raising awareness of parts of speech, word-building and collocations.

Grammatical structures are introduced and reviewed according to the needs of students. Areas such as imperatives and instruction-giving are introduced early in the book to prepare learners for real-life work situations. A whole unit is devoted to numbers, dimensions and quantities, as oil industry workers need to be confident in using these. The passive voice is also given particular emphasis because of the high frequency of passive forms in technical manuals and reports. Discovery learning activities are favoured, where students are encouraged to notice and work out structural patterns rather than learn rules that have little practical application.

Speaking skills development activities simulate real-life communication, such as describing and giving information about equipment and jobs, giving instructions and warnings and discussing workplace problems. These are practised using role-play and information-gap tasks that involve finding out information from a partner.

Reading and listening skills are developed through tasks that involve studying authentic and simplified authentic-type texts, for example toolbox talks and safety leaflets. These are the sorts of texts that oil industry technicians will come across in everyday life. Listening texts prepare students for communication with native speakers and non-native speakers by using recordings that contain different accents and regional variations. Tasks encourage the development of text attack skills that are useful in the real world, such as guessing the meaning of unknown words and scanning a text and picking out key information.

Written tasks aim to motivate students by developing their knowledge of useful language phrases and chunks for different text types such as accident report forms and written notes and instructions. There are also written exercises to reinforce vocabulary and grammatical structures.

There is a strong emphasis on recycling and revision of vocabulary and structures. As well as the review lesson at the end of each unit, the review section at the end of the book reviews and extends material from each unit. These have been written in blocks of three lessons to be used after teaching each section of three units, i.e., Review Units 1–3 are to be used after Unit 3, Review Units 4–6 to be used after Unit 6, and so on.

Procedures

Using the Teacher's notes

The notes for each lesson give step-by-step procedures and answers to exercises as well as a suggested lead-in and closure for the lesson. These procedures can be adapted or added to so that the teacher addresses the needs of the class. Each lesson should incorporate opportunities for students to relate the lesson to their own experiences and needs.

Additional materials and activities

Wherever possible, it is a good idea to supplement the diagrams and pictures in the course by making use of realia (such as genuine equipment and parts), plus additional photographs and illustrations. The Teacher's notes sometimes give links to useful websites

that teachers and/or students can access in order to research the topic in more detail.

Language focus

The course is not designed to focus heavily on grammatical structures and tenses. If teachers wish to spend more time clarifying grammatical areas, they may wish to supplement the course with additional material. On the other hand, some of the written exercises that reinforce language patterns can be omitted, if teachers wish to focus more on transferable skills. Similarly, although many of the exercises and tasks are designed for pairwork, the teacher may prefer to exploit them for individual or group work, depending on the size and nature of the class.

Recycling vocabulary

Additional vocabulary practice and revision can be incorporated into the lesson by using word games and quizzes as in general English classes. The word lists in the Teacher's Book, at the end of each unit, provide a comprehensive list of the principal technical and semi-technical lexical items that occur for the first time in the unit. There is also a complete word list at the back of the book. Students can also be encouraged to keep vocabulary notebooks and to use the glossary at the back of the course book to check meaning and spelling.

Providing extra speaking activities and pronunciation work

Students can be encouraged to look at the tapescripts and read the dialogues aloud, if extra speaking practice is needed. Some tapescripts also serve as a model for pronunciation practice. It is useful to encourage students to listen and identify word and sentence stress as well as weak forms.

Mixed-level groups

Model answers are provided in the Teacher's Book for many of the freer writing or speaking tasks. If there are weaker students in the class, they could be given key words and phrases from the model answers as prompts. Model answers can also be copied onto an OHT during the feedback stage so that students can compare their versions with the model.

GIVING BASIC INFORMATION

Lesson 1: Talking about yourself

Objective

• to practise giving information about yourself and your family

Language

• present simple: *be*; contractions

Vocabulary

• personal and family information

LEAD-IN

• Introduce yourself to students by giving them information about yourself such as your name and job. Then elicit similar information from students and respond appropriately, e.g., *Nice to meet you!*

• If there is a range of jobs in the class, write job titles on the board and check understanding and pronunciation of each title.

A ◆) (CD1 T1) Listen and put sentences in the right order

• Introduce the topic of family members by giving information about your family and asking simple questions about students' families. Review key vocabulary.

• Direct students to the task. Establish that the sentences are not in the right order and that students must write the letters *A* to *H* in the appropriate boxes after they have listened to the recording.

• Play the recording. Give students time to think about and compare their answers with a partner before playing the recording again, if necessary.

• Feed back by eliciting each line of the recording. Ask further questions orally, such as *Is Alan from England? Is Anna from England?*

Answers

1 F 2 H 3 D 4 C 5 B 6 A 7 G 8 E

Tapescript
Presenter:
Unit 1 Giving basic information
Lesson 1 Talking about yourself
A **Listen. What is Alan saying? Put the sentences in the right order, A to H.**

Alan: Hi, my name's Alan. I'm a technical trainer. I'm married with two children. Let me tell you about my family. My wife's name is Anna. My son, Adam, is 13 and my daughter, Sophie, is 10. They're both students. They aren't with me in Azerbaijan.

B Write a biographical paragraph

• Set the task for individual work. When students have finished, ask them to cover the page and spell the words *family*, *son*, *daughter*, *married*, *children*, *technical*, *both*. Write what students spell on the board. Highlight the question *How do you spell ...?*

• Use the paragraph for pronunciation practice if students have problems in this area. Model the sentences and ask students to repeat them.

C Practise contractions of present simple verb *be*

- Elicit what Alan said about his marriage and job (*I'm married. I'm a technical trainer.*). Reinforce the contraction *I'm*. Elicit contractions for the other persons (*you*, *he*, *she*, etc.).
- Ask students to complete the table.
- Feed back on the answers and use this as an opportunity to practise pronunciation of the contractions. Put them into short sentences for repetition, e.g., *He's married. We're from England.*

Answers

Present simple: *be*	
I'm	I am
you're	you are
he's	he is
she's	*she is*
it's	*it is*
we're	we are
they're	they are

D Write sentences about Alan's family

- Ask questions about Alan around the class and check that students are using the correct contractions in their responses.
- Set the exercise for individual work.
- Ask some students to read out their sentences to the class.
- Students may use contractions in the writing exercise. Point out that this is not incorrect in informal texts, but in more formal texts full forms tend to be used.

Example answers

Alan is from Scotland.
He's married with two children.
He's a technical trainer.
His son's name is Adam. He's 13.
His daughter's name is Sophie. She's 10.
They're students.
His wife is from England.
His wife's name is Anna.

E Talk about yourself

- Look at the prompts with students and elicit suitable sentences about the students themselves.
- In pairs, students tell each other some things about themselves and their families.
- Feed back by eliciting some information from individual students about their partner.

F Write three sentences about yourself

- Monitor as students do the task.
- Ask some students to read out their sentences when they have finished.

CLOSURE

- Ask students to write down a sentence with information they learned today about someone else in the class, but they mustn't give the name, e.g., *His wife's name is Munira.*
- Students read out their sentences individually. The class guess who they are talking about.

Lesson 2: Introducing people

Objective

- to practise formal and informal greetings and introductions

Language

- functional language for greetings, introductions and appropriate responses

Vocabulary

- greetings
- introductions

LEAD-IN

- Ask students what details they can remember about you and Alan from the last lesson.

A Find and correct expressions in a dialogue

- Draw attention to the visual and establish the situation. Elicit typical phrases for introducing people.
- Set the task for individual work and pairwork checking.
- Feed back by reading through the dialogue slowly and getting students to stop you when there is a mistake. Do not confirm whether they are correct at this point.

B ⏺ (CD1 T2) Listen to check answers

- Play the recording for students to listen and check their answers.
- Ask students to repeat the exponents to practise the pronunciation. Focus particularly on the intonation of the questions and responses.
- Divide the class into groups of three to practise reading the dialogue.
- Monitor and help where necessary.

- To give extra practice, change the groups and role-play the same situation without looking at the written dialogue. Encourage students to change and develop the dialogue scenario.

Answers

1 *Fine*, thanks.
2 *Do you* know Yusef?
3 Yusef, *this* is Bob.
4 *Pleased* to meet you.
5 No, he's my brother-in-law.

Tapescript
Presenter:
Lesson 2 Introducing people
**B Listen and check your answers.
Then practise reading the dialogue in groups of three.**

Ahmed: Hello, Bob.
Bob: Hi, Ahmed. How are you?
Ahmed: Fine, thanks. And you?
Bob: I'm very well, thanks.
Ahmed: Do you know Yusef?
Bob: No, I don't.
Ahmed: Bob, this is Yusef. Yusef, this is Bob.
Yusef: Pleased to meet you.
Bob: Pleased to meet you, too. Do you work with Ahmed?
Yusef: No, he's my brother-in-law.

C Greet people in different ways

- Ask a student *How are you?* Elicit different responses.
- Set the task for pairwork discussion.
- Feed back to the class. Elicit the difference in formality between *I am very well, thank you* and *I'm all right./Not so bad.* Discuss how it would be appropriate to respond to different people, e.g., the boss of a friend.

Answers

How are you? Not so bad./I'm all right./
I am very well, thank you.
Do you work here too? No, I know Bob from university./No, I work on the seismic crew./Yes, I'm a driller.

D Match greetings, explanations and questions with the responses

- Elicit suggestions for other greetings and introductions, then write them on the board. Elicit any responses.
- Set the task for individual work and pairwork checking before feeding back to the whole class. Compare the exchanges in the exercise with the ideas on the board.
- Practise the greetings and responses as necessary, with books closed.
- Remind students of the formal/informal difference. Elicit which of the pairs of sentences in the exercise use formal or informal language.

Answers

2 Let me introduce my colleague, Vasily.
  Good to meet you.
3 How do you do?
  How do you do?
4 Sorry I'm late. I was held up.
  That's okay.
5 How are your family?
  They're all well, thanks.
6 Haven't we met before?
  I don't think so.
7 Is this your first visit to Azerbaijan?
  No, I was here last year.

8 What have you been up to recently?
  I've been on holiday.

E Write short dialogues

- Ask students to imagine they are meeting their partner for the first time. Ask them to role-play the situation. They can start with some of the expressions from this lesson, then continue in their own words.
- They should then role-play another situation where they are meeting their partner after a long time. Again, they can start with some expressions from the lesson and continue in their own words.
- Each pair can choose which situation to write a dialogue for.
- Monitor students to give help and advice where necessary.
- If there is time, ask one or two pairs to read out their dialogue to the class.

CLOSURE

- Review vocabulary from this and the previous lesson (*I'm not married. I am … . I don't have a brother, but I have a … . My wife's brother is my …,* etc.).

Lesson 3: Asking questions

Objective

- to practise asking questions to find out personal information

Language

- possessive adjectives
- present simple (1st-, 3rd-person singular/plural)

Vocabulary

- question words

LEAD-IN

- Say *I'm very well, thank you* and elicit the question. Repeat with other responses and answers, including exponents from the previous lesson.

A Write personal questions

- Set the task for individual work or pairwork. Go round and monitor for problems, but do not give students the answers at this point.
- They then practise the exchanges with a partner.

B ◀ CD1 T3 Listen to check answers

- Play the recording for students to listen and check their answers.
- In pairs, students ask each other the questions, giving answers that are true for themselves.
- Choose pairs to ask and answer their questions and responses to the class. Correct pronunciation, if necessary.

Answers

1 What's your name?
2 How old are you?
3 Are you married?
4 What's your wife's name?
5 Do you have any children?
6 What are their names?
7 Where do you live?
8 What do you do?

Tapescript
Presenter:
Lesson 3 Asking questions
B Listen and check your answers.

Voice 1: What's your name?
Bob: My name's Bob.
Voice 1: How old are you?
Bob: I'm 32.
Voice 1: Are you married?
Bob: Yes, I am.
Voice 1: What's your wife's name?
Bob: It's Helen.
Voice 1: Do you have any children?
Bob: Yes, we have two.
Voice 1: What are their names?
Bob: Paul and Emma.
Voice 1: Where do you live?
Bob: I live in Azerbaijan.
Voice 1: What do you do?
Bob: I'm a trainee operator.

C Study possessive adjectives and singular/plural verbs

- Review possessive adjectives. Point out something on the table or in the class that is yours. Say *It's **my** bag.* Elicit *his/her/your/their* by indicating other objects. Use the classroom to show *our.* *We are in Room 12. Room 12 is our room.*

- Students complete the task individually and check answers in pairs.
- Conduct whole-class feedback. Select students to read the sections of the text aloud to practise pronunciation.
- Go over the different verb forms. Write *I ... two children* on the board. Elicit *have*. Indicate a student. *Ahmed ... three children.* Elicit *has*. Ask individual students to give you a sentence about themselves and a friend, e.g., *I have one child. Mehmet has two children.* Repeat the activity with the verbs *be*, *work* and *like*.

Answers

Our names are Yusef and Ahmed. I'm 26 and Ahmed *is* 34. We *are* from Algeria. I *am* single. Ahmed *is* married to *my* sister. They *have* three children, and *their* second son is called Yusef, too. We *live* in Algiers and we both *work* on an oil rig. Ahmed *is* a diver, so he *works* under water. I'm a derrick monkey at the top of the rig – I *don't* like going in the water!

D Complete language tables

- Set the task for individual work. Point out to students that they can refer back to the text in exercise C for guidance. Go round and monitor for problems as they do the task.
- Feed back, eliciting answers from the class. Go over any problem areas.
- Give further practice by asking students to discuss in pairs what they have in their bags; where people in their families work and live; what they like.
- They should note down their partner's answers and tell another student in the group.

Answers

Present simple		
Subject	*be*	*have*
I	*am*	*have*
you	*are*	*have*
he/she/it	*is*	*has*
we/they	*are*	*have*

Present simple		
Subject	*work*	*like*
I	*work*	*like*
you	*work*	*like*
he/she/it	*works*	*likes*
we/they	*work*	*like*

Subject	Possessive adjective
I	*my*
you	*your*
he	*his*
she	*her*
it	*its*
we	*our*
they	*their*

E Write a paragraph with more information about yourself

- Give students some information about yourself and a fellow teacher or trainer, using the ideas from the text in exercise C.
- Ask them questions about what you have said.
- Tell students to write a paragraph about themselves and a friend or family member. They can use the text as a model.
- Monitor to help and give advice where necessary.
- Ask one or two students to read out their paragraph.

CLOSURE

- Books closed. Revise the questions from exercise A by asking what questions students could ask Yusef (from the text in exercise C). Encourage students to think of additional questions to ask about Ahmed and write them on the board.
- See if students can remember information about Yusef and Ahmed. Prompt them, if necessary.

Lesson 4: Asking for clarification

Objective

- to practise asking for clarification and meaning

Language

- questions used when asking for clarification and meaning

Vocabulary

- words related to the energy industries

LEAD-IN

- Write some information about yourself on the board, including your family and the place where you live and work. Elicit questions for each piece of information.

- In pairs, students write down some information about an imaginary person and ask each other questions to find out the information. Elicit examples.

A Study ways of asking for clarification

- Say something very quickly to the students. It should be a sentence that they can understand when spoken more slowly. Write students' responses on the board and correct with *Could you speak more slowly, please?* Students repeat the question.

- Go through the same procedure to elicit other questions. Tell students something using a more complex structure than they can understand, say something in a very soft voice, use a difficult word in a sentence. Ask one student the name of his wife or daughter. Indicate that you want to write it on the board, but only get as far as the first letter and elicit the question from the class. Ask them to ask you a difficult question, then ask them some tricky questions to elicit the response.

- Set the task for individual work.

B ◆ (CD1 T4) Listen to check answers

- Play the recording for students to listen and check their answers.

- Conduct feedback and correct any pronunciation mistakes.

- Students practise in pairs by doing the same as you did before when eliciting responses.

Answers

2   Sorry, I don't understand.
3   Could you repeat that, please?
4   What does 'extinguish' mean?
5   How do you spell 'extinguish'?
6   Sorry, I don't know.

Tapescript
Presenter:
Lesson 4: Asking for clarification
B Listen and check your answers.

Voice: 1   Could you speak more slowly, please?
2   Sorry, I don't understand.
3   Could you repeat that, please?
4   What does 'extinguish' mean?
5   How do you spell 'extinguish'?
6   Sorry, I don't know.

C Learn meanings of new words

- Students cover the right-hand column and read the words in the left-hand column. Elicit any meanings they already know. In pairs, they discuss the words to see if they can add to their partners' number of ticks.
- Set the task for individual work and pairwork checking.

Answers

2 to inhale – to breathe in air, smoke or gas
3 to expand – to become bigger in size or number
4 a flame – hot, bright, burning gas that we see when something is on fire
5 flexible – when something can change or move easily
6 a risk – a danger or chance
7 to monitor – to watch or measure an activity
8 to compress – to press something, or make it smaller

D Practise asking for meaning

- Practise pronunciation of new words as a full group.
- Ask one pair of students to read the short dialogue from the speech bubbles aloud. The class repeat.
- Set the task for pairwork.
- Ask individual students questions about the meanings at random.

E Practise the new words in context

- Set the task for individual work and pairwork checking. Elicit the first answer before students start.
- Conduct feedback.
- Check what they can remember by giving the definitions and eliciting the new words. Students test each other in pairs.

Answers

1 It is dangerous to take *risks* on the rig.
2 The metal *expanded* when it was heated.
3 We need to *monitor* the situation carefully.
4 *Compressed* air is used to power drills and motors.
5 Plastic is more *flexible* than concrete.
6 He went to hospital because he *inhaled* some dangerous gas fumes.

F Practise asking for clarification

- Refer students to the visual and elicit what the men are doing and what the warning sign shows. Set the task for individual work and pairwork checking.
- Students write down the correct words.
- Conduct feedback.

Answers

B: Sorry, I *don't* understand.
B: *Could you* repeat *that*, please?
B: *What does* 'inhale' mean?
A: It means *to breathe in air, smoke or gas.*
B: How *do you* spell 'inhale'?

G Practise reading a dialogue

- Model the dialogue for students to repeat. Focus on correct word stress and intonation.
- Students practise reading the dialogue in pairs.
- Ask one or two pairs to read the dialogue for the class.

CLOSURE

- Quickly repeat what you did at the beginning of the lesson to elicit the sentences and questions.
- Ask students to close their books and write down as many new words they learned as they can.
- Elicit examples.

Lesson 5: Making sentences

Objectives

- to identify different parts of speech
- to study basic sentence and question structure

Language

- parts of speech
- sentence and question structure

Vocabulary

- words related to the energy industries

LEAD-IN

- Write *Divers work in deep water* on the board. Point out that every word is a 'part of speech' and these all have a special name. Write the parts of speech above the words as in the course book.

A Categorize different parts of speech

- Ask students to read through the *Parts of speech* information box. Elicit more examples of nouns, verbs and adjectives, and write any correct suggestions on the board. Explain that adjectives tell us more about the noun, e.g., *I live in a house. It is a **big** house.* Verbs usually describe an action, but not always. Sometimes it can be a state like *I **have** two children.*

- Set the task for individual work or pairwork.

- Conduct feedback. For pronunciation practice, students repeat the individual words after your model, with books covered. Reinforce the fact that sound and spelling are not always directly linked.

- In pairs, students link pairs of words from the columns together, e.g., *safe equipment, difficult job.*

Answers

Noun	Verb	Adjective
equipment	learn	exact
operator	expand	difficult
job	monitor	flexible
course	calculate	safe
pipeline	extinguish	dangerous

B Study more parts of speech

- Explain that nouns, verbs and adjectives are the main parts of speech, but that there are others. Write *The floorman always works at the bottom of the rig* on the board. Elicit the parts of speech they can name. Write them over the words. Underline those that are left.

- Set the task for pairwork.

- Ask for answers from different pairs.

Answers

2   conjunctions: *and, but, so* –
    They join one clause to another.
3   adverbs: *well, quickly, sometimes* –
    They give more information about a verb, such as time, place or manner.
4   prepositions: *in, on, from, to* –
    They link nouns to other elements, such as place or time.
5   auxiliary verbs: *have, be, do, will* –
    They are used with a main verb to

change the tense, or form a question.

6   pronouns: *he, you, they, mine* –
    They often replace a noun.

C Identify parts of speech

- Students look again at the example sentence on the board. They identify the underlined words.
- Set the task for individual work and pairwork checking.
- Conduct feedback.

Answers

1   The (article) equipment (noun) is (verb) dirty (adjective) and (conjunction) dangerous (adjective).
2   Bob (noun) speaks (verb) English (noun) well (adverb).
3   The (article) floorman (noun) always (adverb) works (verb) at (preposition) the (article) bottom (noun) of (preposition) the (article) rig (noun).
4   He (pronoun) is (verb) a (article) good (adjective), careful (adjective) worker (noun).
5   The (article) supervisor (noun) makes (verb) sure (adjective) that (conjunction) everyone (noun) follows (verb) the (article) safety (noun) regulations (noun).
6   I (pronoun) go (verb) to (preposition) college (noun) now (adverb), but (conjunction) I (pronoun) will (auxiliary) finish (verb) in (preposition) June (noun).

D Study basic sentence structure

- Write *Bob likes his job* on the board. Elicit that this is a sentence and starts with a capital letter and ends with a full stop. Explain that many English sentences have this structure. *Bob* is the subject because he is doing the liking. *His job* is the object because that is what he likes.
- Set the task for individual work.
- Conduct feedback. Remind students about capital letters and full stops.

Answers

1   Pipelines transport gas.
2   Drillers operate the drill.
3   Derrick monkeys don't work under water.

E Study question forms

- Say *I like Scotland.* Elicit *Do you like Scotland?* Write the sentences on the board. Do the same with *Aberdeen is in Scotland. Where is Aberdeen?*
- Elicit the subject, verb and object in question 1. Remind students that *do* is an auxiliary verb for questions and negatives. Elicit the verb and subject in question 2. W*here* is a question word and we do not use *do/does* in this question because it is the verb *be*. With the verb *be*, we invert verb and subject.
- Set the task for individual work.
- Conduct feedback.

Answer

before

F Rearrange words to form questions

- Set the task for individual work. Conduct feedback. Students name the subject, verb, object (and question word).
- Point out that *Wh-* questions usually have falling intonation, whereas *Yes/No* questions have rising intonation.
- Model the intonation of the questions.

Answers

1   Do you work on a rig?
2   Where are my overalls?
3   What does this word mean?

CLOSURE

- Test students with definitions of the new vocabulary from exercise A.
- Students test each other on parts of speech by writing down ten different words and swapping lists. They identify the parts of speech.

Lesson 6: Identifying equipment

Objectives

- to learn common words related to the oil industry
- to learn how to spell common oil industry words

Language

- spelling words

Vocabulary

- equipment and devices

LEAD-IN

- Start with a quick revision of the previous lesson. Ask for examples of a noun, a verb, an adjective, an article, an adverb, an auxiliary verb and a preposition.
- Give definitions to elicit vocabulary, e.g., *pipeline*, *flexible*, *expand*, *calculate*, *monitor*, *safe*, *equipment*, *operator*.

A Study new vocabulary related to equipment and devices

- Ask students to name one item of equipment they use in their jobs. Write them on the board.
- Use items in the classroom to show how to use *this/that* and *these/those*.
- Books open. Students read the speech bubbles.
- Set the task for pairwork. Do the first one with them. Ask *What's this?* Elicit *It's a chain saw.*
- Students continue to ask and answer in pairs.
- Check answers by naming a pair and giving a number. That pair asks and answers the question about it.
- Ask questions quickly around the class, e.g., *What's number 6? What's number 9?* Go over any pronunciation problems.

- Ask students which (if any) items they use in their jobs.

Answers

1   chain saw
2   socket
3   bolts
4   hard hats
5   rig
6   crane
7   pipeline
8   plug
9   gauge
10 pig
11 forklift
12 overalls

B Pronounce letters correctly

- Check students know the English alphabet. Dictate some letters to the students, e.g., *A, E, G, I, J, U, W, X, Y, Z*. Monitor to make sure they write them correctly.
- Refer students to the table and check they understand the different sounds shown.
- Set the task for individual work. Students then compare answers with a partner. Do not confirm whether they are correct at this point.

C 🔊 (CD1 T5) **Listen to check answers**

- Play the recording for students to listen and check their answers.
- Students read the letters from the groups with correct pronunciation.
- Go through the alphabet again from A to Z. Ask students to write the alphabet backwards, in pairs. The pair that finishes first has to read it out to the class.

Answers

/eɪ/ (play)	/iː/ (see)	/e/ (men)	/aɪ/ (my)	/əʊ/ (go)	/ɑː/ (car)	/uː/ (do)
J K	B C D E G P T V	F L M N S X Z	I Y	O	R	Q U W

Tapescript
Presenter:
Lesson 6: Identifying equipment
C **Listen and check your answers.**

Voice: /eɪ/ play – A – H – J – K
/iː/ see – B – C – D – E – G – P – T – V
/e/ men – F – L – M – N – S – X – Z
/aɪ/ my – I – Y
/əʊ/ go – O
/ɑː/ car – R
/uː/ do – Q – U – W

D **Practise spelling**

- Ask individual students to spell *inhale* and *flexible*. Write what they say on the board. They must be clear.
- Set the task for pairwork. Before students start, run through the correct pronunciation of the words again. This time they can look at the words as they spell them.

- Books closed. Ask individual students to spell words from the exercise from memory.
- Students work in pairs again to test each other on the spellings. One student has their book open to check the spelling, the other has to spell from memory.

E **Correct spellings**

- Set the task for individual work and pairwork checking.
- Conduct feedback.

Answers

1 socket
2 forty; overalls
3 operator
4 difficult; instrument

CLOSURE

- If students need more spelling practice, ask them to look back through Lessons 1 to 5 and find ten words to test their partners' spelling. They should write down the words in a list. In pairs, they take turns to ask their partner to spell them aloud.

Lesson 7: Talking about opposites

Objective

- to practise using adjectives and their opposites

Language

- position of adjectives

Vocabulary

- adjectives related to the energy industries

LEAD-IN

- Elicit common opposites such as *hot* and *cold*. Use them to compare two things, e.g., *In this country, it's always very hot, but in countries like the UK, it's sometimes very cold.*

- Ask for more common opposites students know, perhaps using the words on the board.

Answers

2 smooth
3 shallow
4 deep
5 approximate
6 accurate
7 flexible
8 rigid
9 contract
10 expand

A Make opposites

- Refer students to the pairs of pictures without looking at the words in the box.

- Explain that the pictures show opposites. Elicit any words they can guess.

- Set the task for individual work and pairwork checking.

- Conduct feedback, checking correct pronunciation.

- Ask for examples of nouns that can be used with these adjectives and verbs, e.g., *deep sea*, *contract – metal*.

B Practise new words

- Set the task for individual work and pairwork checking.

- Feed back by asking individual students for full sentences, not just the word.

- Model the sentences for students to repeat.

- If necessary, ask students how to spell the words they used, with their books covered.

Answers

1 Iron is a *rigid* material.
2 When you heat metal, it *expands*.
3 It is important to be *accurate* when measuring.
4 A 20-metre well is *shallow*.
5 Glass is *smooth*.

C Practise writing sentences using opposites

- Elicit which five words were not used in exercise B.
- Ask students to create example sentences with these. You could make this more interesting by asking them to create a gap-fill exercise for their partners.
- Feed back examples of students' sentences for the rest of the class to complete orally.

D Study more adjectives and their opposites

- Refer students to the speech bubbles and ask them to repeat the short dialogue after your model.
- Set the task for pairwork. This is intended to test students on adjectives they are likely to know already. If there are any they are unsure about, go over them during feedback.

E Write opposites of adjectives

- Set the task for individual work and pairwork checking.
- Conduct feedback and ask for spellings. Focus on the spelling of words with double letters and endings, e.g., *ful/less*.
- For reinforcement, ask students if they can think of five things they can describe with five of the adjectives. They should say the noun and their partners must guess the adjective, e.g.,
  A: *My car.*
  B: *Big!*
  A: *My job.*
  B: *Easy!*
- Go through the adjectives and elicit a noun for each.

Answers

A
fast – slow
long – short
early – late
careful – careless
big – small
cool – warm
difficult – easy
flammable – inflammable
sad – happy

B
good – bad
dark – light
loud – quiet
dirty – clean
complex – simple
dry – wet
hot – cold
dangerous – safe
wrong – right
fragile – strong

CLOSURE

- Give a quick oral test of opposites covered in the lesson. Give an adjective and elicit the opposite from the class.
- In pairs, students do the same with a partner. They take it in turns to look at the book and ask for the opposites.

Lesson 8: Talking about shapes and sizes

Objective

- to learn and practise vocabulary related to shapes and sizes

Language

- position of nouns and adjectives

Vocabulary

- shape nouns and adjectives
- size adjectives

LEAD-IN

- Get students to draw different shapes on paper and compare them with a partner.
- Ask for examples. Students draw in the air, give you the paper or draw on the board. Elicit the names of the shapes they know.

A Study words to describe shapes

- Students compare the number of shapes in their books and on the board.
- Point out the different columns: *Shape*, *Noun* and *Adjective*. Remind students of the difference.
- Set the task for pairwork. Compare and add answers with another pair, if necessary.
- Point out shifting word stress in *triangle/triangular* and *rectangle/rectangular*. Do the same for *cylinder* and *hexagon*.
- Feed back answers from pairs.

Answers

Shape	Noun	Adjective
1	a triangle	*triangular*
2	a rectangle	*rectangular*
3	*a square*	square
4	a circle	*circular*
5	*a hexagon*	*hexagonal*
6	*a cylinder*	cylindrical
7	*a sphere*	*spherical*
8	*a cube*	cuboid

B (CD1 T6) Listen to check answers

Tapescript
Presenter:
Lesson 8: Talking about shapes and sizes

B Listen and check the pronunciation of your answers.

Voice:		
1	a triangle	triangular
2	a rectangle	rectangular
3	a square	square
4	a circle	circular
5	a hexagon	hexagonal
6	a cylinder	cylindrical
7	a sphere	spherical
8	a cube	cuboid

C Practise talking about shapes

- Choose something in the classroom that is a particular shape. Tell students you can see something that is *round*, *square*, etc. Ask them to guess what shape you are thinking about.

- Put students into small groups to make a list of things they can see that are different shapes.

- In an open group they guess what is on the other groups' lists.

- Elicit examples of things that are different shapes in the students' workplaces or that they use in their work.

D Review opposites

- Set the task for individual work and pairwork checking.

- Feed back answers from the group.

- Elicit examples of common nouns that these adjectives can describe, either in the room or from the students' homes, e.g., *thick/thin walls*, *deep/shallow swimming pool*, *long/short path/shelf*, *high/low wall/table/desk*, *big/small board*.

Answers

2 wide – narrow
3 deep – shallow
4 long – short
5 high – low
6 thick – thin

E Practise using adjectives and nouns

- Refer students to the visuals and ask them to guess what they represent. Clarify any words or diagrams that are not clear, then set the task for individual work and pairwork checking.

- Look at the example in the task and elicit the opposite.

- Check answers by asking individual students.

- Give students some pronunciation practice of reading two words together. Point out that, in English, we often run words together when we pronounce them in connected speech.

- Say some of the pairs of words quite quickly and naturally and ask students to identify which pairs you said. Students practise repeating them.

Answers

2   a small *hole*
3   a *narrow* ring
4   a *wide* ring
5   low *pressure*
6   *high pressure*
7   *a small explosion*
8   *a big explosion*
9   *a short pipe*
10 *a long pipe*

CLOSURE

- Ask students to think of a noun in the energy industries that has a particular shape and can be described using an adjective from exercise D. Their partners must guess what it is.

Lesson 9: Describing things

Objective

• to learn and practise language used to describe and define things

Language

• position of adjectives

Vocabulary

• colours
• adjectives
• words related to rigs

LEAD-IN

• Elicit different coloured items in the classroom. Encourage students to put colours and the items together to practise pronunciation, e.g., *There's a red pen. There's a black computer screen.* Students should be familiar with colours, so make this brief.

A Use colour adjectives to describe objects

• Elicit examples of different coloured items in the students' workplaces.

• Set as pairwork. Make this a race by seeing who can complete their list the fastest.

B Practise using colour adjectives

• Focus on the fact that we can have more than one adjective before a noun, e.g., *I have a big, red car.*

• Set the task for pairwork.

• Conduct feedback by asking different pairs for their sentences.

• Elicit where the colour adjective usually goes when there are two adjectives (in second place).

• Refer students to the visual and ask them which of the things from the exercise they can see (*barge*, *pipe* and *crane*). Elicit other things that can be seen in the visual.

• For extra practice, ask students to write three phrases of their own like those in exercise B, using two adjectives before a noun. They could mix up the letters of the adjectives for their partners to unjumble in order to make this more fun and communicative.

Answers

2 a *long thin* cable
3 a *tall grey* crane
4 *dirty orange* overalls
5 a *thick white* pipe

In the picture, you can see a large barge, a long thin cable and a tall grey crane.

C Label a diagram

• Write the words *terminal*, *wire* and *connect* on the board and elicit where you find all three. Refer students to the diagram and elicit which are the terminals and which are the wires. Point out the key on the right of the diagram which shows the different colours.

- Set the task for individual work and pairwork checking.
- Feed back answers from the whole group.
- Ask questions for students to answer, using only the diagram, not the sentences, e.g., *Which wire connects terminal A to terminal B? Which wire connects terminal B to terminal D? Which wire connects terminal A to terminal C? Which wire connects terminal C to terminal D?*
- Students ask the same questions in their pairs to practise vocabulary and pronunciation, if necessary.

Answers

2   The green wire connects terminal B to terminal D.
3   The brown wire connects terminal A to terminal C.

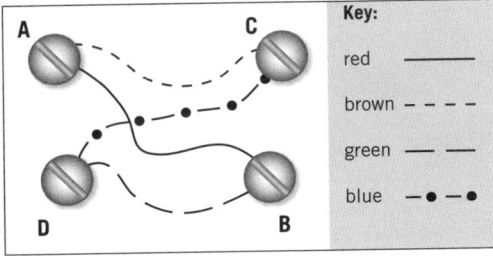

D Read definitions

- Give an example of a definition, e.g., *something you wear to protect your head when you are working – a hard hat.*
- Set the task for pairwork. Explain that it is not important to understand every word. Students should try to guess from what they do understand.
- Check answers by asking different pairs. Students spell the word they give you clearly so that you can write it on the board.
- Ask students to find which words are in the glossary and compare the definitions. Point out that the glossary is a useful tool that students will use more as they work through the book.

Answers

2   derrick
3   rig
4   geophones
5   pontoon

E Correct definitions

- Write an incorrect definition on the board, e.g., *a flexible, plastic object used for writing: a pen.* Elicit what the mistake is.
- Set the task for pairwork.
- Conduct feedback from the whole group.

Answers

2   not usually pink
3   thin and flexible, not thick and rigid
4   thick and black, not thin and blue

F Write definitions

- Set the task for pairwork. Encourage students to use the sentences in exercise E as models.
- Conduct feedback from the whole group.

CLOSURE

- Books closed. Give students a quick test on the lesson.
- Ask them to write down as many colours as they can.
- Ask them what connects two terminals.
- Ask them to name five things from this lesson beginning with *m*, *d*, *r*, *g* and *p*.

Lesson 10: Giving definitions

Objectives

- to learn job titles
- to identify different types of rigs from descriptions

Vocabulary

- job titles for oil rig workers

LEAD-IN

- Students work in pairs. Ask them to write down as many different jobs connected with their work as they know.

- Ask for examples. Discuss what jobs students do and which jobs they have had or would like to do.

A 🔊 (CD1 T7) Match jobs and definitions

- Refer students to the list of jobs. Read through the list of definitions and see if students know which job is being defined. Then read through the list of jobs to model pronunciation.

- Set the task for individual work and pairwork checking. Encourage students to use the alphabetical glossary at the back of the course book to find definitions of oil industry jobs.

- Play the recording for students to listen and check their answers.

- Ask students if their job is included in the list. If not, ask how they would define it. Discuss whether the students have ever had any of these jobs and if they enjoyed them. To reinforce the vocabulary, ask *Who* questions, e.g., *Who assists the driller?* etc.

Answers

2   A derrick monkey works at the top of the derrick.

3   A doodlebugger works on the seismic crew.
4   Drillers operate the drilling machinery.
5   A floorman works on the floor of the derrick.
6   A jughustler uses geophones on the seismic crew.
7   A metallurgist studies rocks to find oil deposits.
8   Motormen control the drilling engine.
9   A mudman maintains the mud systems.
10   A roughneck assists the driller.
11   Roustabouts do routine cleaning and maintenance.

Tapescript
Presenter:
Lesson 10: Giving definitions
A   Match each job with its definition. Then listen and check your answers.

Voice: 1   Divers work under water.
2   A derrick monkey works at the top of the derrick.
3   A doodlebugger works on the seismic crew.
4   Drillers operate the drilling machinery.
5   A floorman works on the floor of the derrick.
6   A jughustler uses geophones on the seismic crew.
7   A metallurgist studies rocks to find oil deposits.
8   Motormen control the drilling engine.

9 A mudman maintains the mud systems.
10 A roughneck assists the driller.
11 Roustabouts do routine cleaning and maintenance.

B Label a diagram with places where people work

- Refer students to the visual and elicit what sort of platform this is.
- Set the task for pairwork.
- Conduct feedback by drawing an outline of the platform on the board and asking students to indicate the different places. Accept different answers and ideas.

C Identify different types of rigs

- Elicit why different types of rig are used and not only one (because of different conditions; how deep the water is). Elicit names of different rigs that the students know.
- Set the task for individual work and pairwork checking. Explain that students do not need to know all the words to do the task.
- Elicit feedback from the whole group and ask for reasons for their choice of answers.
- Check the meaning of problem vocabulary. Point out that there are short definitions of all the rigs in the glossary.

Answers

a 2 Barge-type
d 4 Tension leg platform
b 5 Jack-up
c 1 Fixed platform
e 3 Semi-submersible

CLOSURE

- Give students some time to look again at the new words in the lesson, then give a spelling test. Test some of the jobs and the different types of rig.
- Check the answers by asking individual students to spell the words for you to write on the board. When they give you the word, ask for a definition.

Answers

A

A	R	T	D	S	P	W	C	D	G	H	K
M	T	H	E	O	R	E	T	E	C	A	L
R	Y	P	L	A	T	F	O	R	M	R	W
O	Z	S	E	A	N	E	R	R	U	D	S
U	E	X	T	I	N	G	U	I	S	H	E
S	X	A	I	D	I	B	T	C	H	A	I
T	P	L	Y	F	P	N	W	K	J	T	S
A	L	X	B	H	M	U	D	A	E	N	M
B	O	S	P	F	L	W	B	T	N	S	I
O	S	H	P	I	E	N	T	E	M	O	C
U	I	A	N	X	I	D	C	R	A	N	E
T	O	L	O	E	A	I	N	D	B	S	C
R	N	L	N	D	M	O	N	I	T	O	R
A	B	O	I	B	J	E	C	T	I	C	E
B	D	W	T	B	N	Z	T	S	L	K	R
S	K	C	O	N	T	R	A	C	T	E	R
A	F	I	R	D	A	H	L	L	Y	T	N
N	S	H	E	X	A	G	O	N	A	L	D

B

1 shallow
2 hard hat
3 monitor
4 roustabout/roughneck
5 hexagonal
6 contract

C

John is a roughneck on an oil rig. He works *on the floor* of the rig and helps the driller operate the *drilling machinery*. He wears a hard hat and *overalls*. The rig is a semi-submersible rig that is situated in *deep* waters. It is supported by a *pontoon*.

 CD1 T8

Voice: John is a roughneck on an oil rig. He works on the floor of the rig and helps the driller operate the drilling machinery. He wears a hard hat and overalls. The rig is a semi-submersible rig that is situated in deep waters. It is supported by a pontoon.

Word list

accurate
approximate
barge
bolt
cable
calculate
chain saw
circular
contract
crane
crew
cuboid
cylindrical
derrick
diver
doodlebugger
driller
equipment
exact
expand
explosion
extinguish
fixed
flame
flammable
flexible
floorman
forklift
fragile
gauge
geophones
hard hat
hexagonal
hole
inhale

jack-up
jughustler
metallurgist
monitor
motorman
mud
mudman
operator
overalls
pig
pipeline
platform
plug
pontoon
pressure
rectangular
rig
rigid
ring
risk
rough
roughneck
roustabout
seismic
semi-submersible
shallow
smooth
socket
soiled
spherical
tension
terminal
trainee
triangular

CALCULATING AND MEASURING

Lesson 1: Saying numbers

Objective

• to say and write numbers correctly

Vocabulary

• types of numbers (telephone numbers, dates and quantities)

LEAD-IN

• Elicit different numbers by asking how many people work in the students' workplaces and their companies. Ask if they know the number of people who work in their company and the population of their town or city. Get students to write these down as numbers.

• Conduct feedback. Write the examples.

A Say numbers

• Write the year you were born or an important year for you on the board. Ask students to read the number. Inform them it is an important number for you. Write more examples on the board of numbers for different things. Elicit what they refer to.

• Set the task for pairwork.

• Conduct feedback. Ask individual students to say the numbers. Correct mistakes, but do not go into detail at this point.

• Elicit what students think the numbers refer to. Do not confirm whether they are correct at this point.

B 🔊 (CD1 T9) Listen to check answers

• Play the recording for students to listen and check pronunciation. Stop the recording after each number to allow students to repeat for practice. Draw attention to the use of the weak form of *and*, e.g., *three hundred and eighty-nine*.

• Say that telephone numbers may use the word *zero* as an alternative to *oh*. Also, we say *two thousand and six* for dates at the beginning of a century (*2006*), but for later dates we divide the figure in two, e.g., *twenty twenty*. These will be covered in Lesson 2.

Answers

1 the thirty-first of December, 2006: That's a date.
2 twenty twenty: That's a year.
3 one thousand, three hundred and eighty-nine: That's a number.
4 thirty-three point three percent: That's a percentage.
5 double oh, double four, two, oh, eight, two, five, double oh, four, oh, three: That's an international phone number.
6 ten sixty-six: That's a year.
7 oh, double seven, five, one, seven, six, three, two, nine, eight: That's a mobile phone number.
8 a million pounds: That's a sum of money.
9 eleven point three five: That's a number in decimals.
10 two hundred and thirty-one square metres/metres squared: That's an area.

Tapescript
Presenter:
Unit 2 Calculating and measuring
Lesson 1 Saying numbers
B Listen and check your pronunciation of the numbers.

Voice: 1 the thirty-first of December, two thousand and six

2 twenty twenty
3 one thousand, three hundred and eighty-nine
4 thirty-three point three percent
5 double oh, double four, two, oh, eight, two, five, double oh, four, oh, three
6 ten sixty-six
7 oh, double seven, five, one, seven, six, three, two, nine, eight
8 a million pounds
9 eleven point three five
10 two hundred and thirty-one square metres

C Say and write numbers as words

- Write some high numbers on the board, e.g., *100, 149, 298, 1,000, 1,567, 12,306, 1,000,000, 2,875,140.* Ask students to say them. Ensure they use *hundred, thousand* and *million* in the singular, and *two hundred and fifty*, not *two hundreds and fifty*.

- Ask students to write one or two of the numbers as words. Check spelling and hyphenation, e.g., *forty-nine*.

- Set the task for individual work. Point out that the commas and dashes should help them.

- Conduct feedback. Ask students to spell out the first few answers for you to write on the board. Then write the rest of the answers yourself on the board for them to read and check. Ask for examples of mistakes they made.

Answers

1 one
2 twelve
3 a/one hundred and twenty-three
4 one thousand, two hundred and thirty-four
5 twelve thousand, three hundred and forty-five
6 one hundred and twenty-three thousand, four hundred and fifty-six
7 one million, two hundred and thirty-four thousand, five hundred and sixty-seven
8 twelve million, three hundred and forty-five thousand, six hundred and seventy-eight

D Practise listening to, saying and writing numbers

- Set the task for pairwork. Students should write the numbers rather than the words. They should not let their partners see the numbers they are dictating.

- Conduct feedback.

Answers

Student 1's numbers:
1 12 2 2,143 3 321,654 4 87,645,231
5 1,243,657,890 6 2,060

Student 2's numbers:
1 11 2 132 3 12,354 4 4,213,756
5 123,456,789 6 1,050

E Practise using numbers

- Students do a quick quiz with numbers. All the questions ask *How many?*

- Set the task for individual work. Students should try to answer the questions as quickly as they can, then ask their partners for the answers they don't know or are unsure of.

- Feed back answers around the class. Students guess any they are unsure of.

- For extra practice, in pairs, students make five more quiz questions about numbers to give another pair to answer.

Answers

2 116 3 100 4 3,600 5 12 6 8 7 5 8 2
9 21 10 453.592

CLOSURE

- Write some numbers on the board connected with the college or building you are in. These should be estimates, e.g., the number of people who work here, the number of rooms, floors, lifts, rest rooms, stairs and windows.

- Students guess what they refer to. Do they agree with your estimate?

- Ask them to change the numbers according to what they think.

Lesson 2: Talking about dates and times

Objective

• to say and write times, dates and telephone numbers

LEAD-IN

• Review numbers from the previous lesson.

A Tell the time

• Ask students what the time is and write it on the board. Elicit different ways to ask the time, e.g., *What's the time, please? Have you got the time, please?*

• Elicit the difference between *UK standard time* and *International time*.

• Set the task for individual work.

• Conduct feedback. Do not confirm whether they are correct at this point.

B ● CD1 T10 Listen to check answers

• Play the recording for students to listen and check their answers.

Answers

UK standard time		International time
twelve o'clock	12.00	twelve o'clock
ten past twelve	*12.10*	twelve ten
quarter past twelve	12.15	*twelve fifteen*
twenty-five past twelve	12.25	twelve twenty-five
half past twelve	*12.30*	twelve thirty
twenty to one	12.40	*twelve forty*
quarter to one	*12.45*	twelve forty-five
ten to one	12.50	*twelve fifty*

Tapescript
Presenter:
Lesson 2 Talking about dates and times
B Listen and check your answers.

Voice 1:	twelve o'clock
Voice 2:	twelve o'clock
Voice 1:	ten past twelve
Voice 2:	twelve ten
Voice 1:	quarter past twelve
Voice 2:	twelve fifteen
Voice 1:	twenty-five past twelve
Voice 2:	twelve twenty-five
Voice 1:	half past twelve
Voice 2:	twelve thirty
Voice 1:	twenty to one
Voice 2:	twelve forty
Voice 1:	quarter to one
Voice 2:	twelve forty-five
Voice 1:	ten to one
Voice 2:	twelve fifty

C Practise asking for and giving the time

• Write some times on the board for students to practise saying the time.

• Point at a time and say either *UK standard* or *International* before students answer.

• Write on the board some times that are relevant to you during the day and elicit what students think you do at these times, e.g., the time you get up, etc.

• Refer students to the words in the word box. They can add more verbs to the list.

• Set the task for pairwork.

• Conduct feedback. Ask students to report information about their partners using *he/she*.

D 🔊 (CD1 T11) **Listen to identify times**

- Set the task for individual work and pairwork checking.

- Play the recording. Students listen to write down the times.

- Conduct feedback from the whole group. Extend by putting the times on the board and eliciting what the speaker does at these times. Check if students remember how the speaker said *not exactly* (*just before 8, usually around 7.55*).

Answers

6.30	(He wakes up.)
8.00	(He starts work.)
7.15	(The bus collects him.)
7.55	(He arrives at the terminal.)
12.15	(He breaks for lunch.)
13.15	(He starts the afternoon shift.)
16.50	(He finishes work.)
17.00	(The bus collects him.)
18.10	(He gets home.)

Tapescript
Presenter:
D **Listen and write the times you hear.**

Voice: Well, I wake up at six thirty every morning because I start work at eight. The bus collects me at seven fifteen and we get to the terminal just before eight, usually around seven fifty-five. We break for lunch at twelve fifteen and start the afternoon shift at one fifteen. We finish at four fifty because the bus collects us at five. It takes longer to get into the city because of traffic, so I usually get home around six ten.

E Say phone numbers, years, dates

- Remind students that we can say different sorts of numbers in different ways. Write a mobile phone number on the board including zero and some double numbers. Ask students to read it. Point out that we read *0* as *oh* or *zero*. If there are two numbers the same, we can say *double* (*seven*). Wipe off the number and see how many students can remember the full number.

- Students test each other in pairs by writing down numbers of increasing length for the other to read, remember and say. They repeat this by only saying the number. Ask students to look at the first box and choose the correct option to complete the rule.

- Elicit the years students were born and put some on the board. Practise reading the years with students. Add some more years such as *1910* and *2006*. Students choose the correct option to complete the rule in the second box.

- Elicit birthdays and go through the way we say dates. Model the pronunciation for students to repeat and point out the weak form *of*. Students choose the correct option to complete the rule in the third box.

Answers

Telephone numbers:
Say the numbers *individually*.
Years:
Say the numbers *in pairs*.
Dates:
Say the *day*, the name of the month, then the *year*.

NB: Americans say the month, then the day, then the year. This can cause confusion.

F Practise asking and answering questions about different numbers

- Demonstrate the task by writing a date that is important for you on the board. Elicit the question *Why is that date important?* Respond appropriately.

- Set the task for pairwork.

- Conduct feedback. Ask individual students to tell the class one of their partner's important dates, and the reason.

CLOSURE

- Books closed. Ask for the two ways of telling the time and important dates.

Lesson 3: Talking about fractions and percentages

Objective

• write and say fractions and percentages as words and numbers

Vocabulary

• fractions and percentages
• words related to injuries at work

LEAD-IN

• Review the previous lessons by writing a phone number, a year, a time, a date and an amount on the board and eliciting what they are. Then add a fraction and a percentage and elicit what these are called too.

A Say and write fractions and percentages

• Ask how many students went to bed before or after midnight last night and ask for calculations of this number as a fraction and a percentage.

• Elicit how students travelled here today or whether they spent their last holiday in this country or abroad, and calculate the fractions and percentages related to these numbers.

• Set the task for individual work and pairwork checking.

• Conduct feedback. Ask students to read the numbers they have written down. Spell the more difficult words (*half, quarter, third, eighth, fifth*). Do not confirm answers at this point.

• Practise pronunciation of the numbers. Pay particular attention to the plurals (*thirds/eighths*). Point out that we say *twenty percent*, not *twenty percentage* or *twenty percents*. When there is a decimal point, we say, e.g., *three point five percent*.

B Match fractions and percentages

• Set the task for individual work.

• Conduct feedback with the whole group.

Answers

2 three quarters – seventy-five percent
3 a third – thirty-three point three percent
4 a fifth – twenty percent
5 seven eighths – eighty-seven point five percent

C Practise using fractions and percentages in discussion

• Set the task for pair discussion. Check the meaning of *proportion* and *unemployed*.

• Feed back answers and find out which student in the class spends the longest percentage of his or her time sleeping or working. Also ask students if they think the proportion of children learning English is good.

• You could ask students to think of some more questions that the other students could answer using fractions and percentages.

D Read a text for specific information

• Write the words *health and safety* on the board and elicit what they refer to. Elicit the types of injuries can that happen at work, and what can cause these accidents. Ask what causes most accidents where students work. Pre-teach or check *injuries, slips* and *trips* by demonstration and eliciting.

- Set the task for individual work and pairwork checking.
- Conduct feedback with the whole group.
- Ask students if they think the percentages in the table apply to their workplaces too. How would they adjust them?

Answers

Cause of injury	Percentage
slips, trips or falls	36%
driving related	25%
using faulty equipment	12%
not using appropriate PPE	11%
using the wrong equipment	8%
falling objects	8%

E ◄》 (CD1 T12) **Listen for detailed information**

- Refer students to the visual and elicit *pie chart*. Students read through the key. Elicit what the pie chart is of.
- Elicit or check meanings of parts of the body.
- Set the listening task for individual work and pairwork checking. Play the recording for students to listen for the information.
- Conduct feedback with the whole group.

Answers

Finger or thumb: 16% **Hand:** 6%
Leg or lower limb: 9% **Other:** 3%

Tapescript
Presenter:
Lesson 3 Talking about fractions and percentages
E **Listen and finish labelling the pie chart.**

Voice: Most injuries in the workplace are back injuries. These account for forty-five percent of all injuries. Thirteen percent of injuries are to the arm, and six percent

are to the hand, but even more injuries involve fingers and thumbs – sixteen percent, in fact. Leg or lower limb injuries are less common and account for nine percent of all injuries. Eleven percent of injuries are to other parts of the body – that's eight percent to the torso, and three percent to other areas.

F Discuss a pie chart

- Refer students to the next chart and key and elicit what the chart shows.
- Write on the board *100 accidents in total. 8 accidents. 34 accidents.* Read through the three comments in the speech bubbles and elicit which numbers they relate to.
- Point out that there are three different ways we can talk about the reason for something: *involved, be caused by, be due to.*
- Model the comments for the students to repeat for practice and substitute different types of accidents so that they give full sentences, e.g., *over a quarter of all accidents involved trips and falls*, etc.
- Set the task for pairwork.
- Conduct feedback. Ask for examples from different pairs.

CLOSURE

- If there is time, you could ask students to work in pairs to write questions for a short survey. They are going to ask the rest of the students the questions and note down their answers.
- First, they must choose the topic for the survey. Elicit examples of topics: *leisure activities, daily routine*, etc.
- Students write five questions and circulate to ask the other students their questions, noting down answers. When they have all the answers they need, they draw a pie chart to show the percentages and fractions. From this, they should make sentences about the pie chart.
- Conduct feedback on the results with the whole group.

Lesson 4: Talking about nouns

Objective

• to study countable/uncountable nouns and words used with them

Language

• the use of countable/uncountable nouns and quantifiers

Vocabulary

• countable/uncountable nouns related to the energy industries

LEAD-IN

• Review the reasons for many accidents at work. As students mention the words, write the following on the board: *accident – accidents, fall – falls, machinery – equipment, object – objects.*

• Indicate the words on the board and point out that we can say *one accident – two accidents.* Elicit whether we say the same with *machinery* and *equipment.* Explain that there are two types of nouns, *countable* and *uncountable*, because one type we can count and the other we can't.

A Categorize countable and uncountable nouns

• Elicit examples of countable and uncountable nouns that students know. For uncountables, elicit liquids, gases and materials. Draw a glass of water and a tap with water gushing out. Elicit that one we can count (a glass of water), but the other we can't.

• Set the categorizing task for pairwork.

• Conduct feedback with the whole group.

Answers

Countable nouns: pipeline; tank; filter; engineer; report; engine
Uncountable nouns: gas; sand; water; equipment; power; zinc

B Add plural forms

• Ask students to add plural forms to the table wherever possible.

Answers

pipelines; tanks; filters; engineers; reports; engines

C Study words used with countable and uncountable nouns

• Ask students to work in pairs to make sentences with *How much* and *How many* from the table in exercise A.

• Choose examples from the students and write them on the board: *How many engineers are on the rig? How much water is in the bottle?* After *engineers*, write *a 2; b 52.* After *water*, draw two bottles, one nearly full and one nearly empty. Elicit the questions and answers with *a lot/a little.*

• Set the task for individual work.

• Conduct feedback with the whole group.

Answers

1 some
2 a lot of work
3 There is
4 How many
5 There aren't enough
6 A few

D Study more quantifiers used with countable and uncountable nouns

- Set the task for pairwork. If students are not sure which words are used with which, they should make a sentence with them and see if it sounds strange or not.
- Conduct feedback with the whole group.

Answers

Countable nouns: delete *a little*
Uncountable nouns: delete *many*; *a few*

E Discuss and note down uses of quantifiers

- Refer students back to the italicized words in exercise C.
- Ask students to go through them and discuss which quantifiers are used with countable and which are used with uncountable nouns (and whether they are used with singular or plural nouns if countable).
- They can use the writing lines to make notes. Write the answer to number 1 on the board as an example. Students could also write their own example sentences for reinforcement.
- Conduct feedback with the whole group. Point out that *a lot of* tends to be used in spoken English for affirmative sentences, and *much* and *many* are used in negative sentences and questions, although *many* is often used in formal, written affirmative sentences, e.g., *Many oil rigs will be affected by the new regulations.*

Answers

1   an – countable (singular); some – both (plural)
2   a lot of – both (plural)
3   There are – countable (plural); There is – both (singular)
4   How many – countable; How much – uncountable

5   There aren't enough – countable (plural); There isn't enough – uncountable
6   A few – countable (plural); A little – uncountable

F Identify incorrect sentences

- Write an incorrect sentence on the board, e.g., *There is a lot of accidents in the oil and gas industry.* Elicit that it is wrong, and why.
- Set the task for pairwork. Demonstrate how the activity works with a strong student. Make sure students are looking at the relevant pages.
- As students do the activity, monitor and pick up on any problem areas to discuss with the whole group. There is no need to have class feedback on all the answers.

G Complete sentences with the best options

- Set the task for pairwork.
- Conduct feedback. Ask different pairs for their answers and also their reasons for their choices.

NB: Many of the answers depend on the individual student and/or his or her company.

Example answers

1   All
2   Some
3   A few
5   A lot of (or *a little*, depending on where students compare it with)

CLOSURE

- Review nouns used in the lesson. Students say whether they are countable or uncountable.

Lesson 5: Talking about units of measurement

Objective

• to say and use terms for different units of measurement

Vocabulary

• words related to units of measurement

LEAD-IN

• Write the word *measurement* on the board. Elicit what things we can measure.

A Categorize units of measurement

• Set the task for pairwork.
• Conduct feedback with the whole group. Elicit when students use these different ways of measuring in their work, and what sort of equipment they can use to do the measuring.

Answers

1 Length 2 Time 3 Electrical 4 Weight
5 Pressure 6 Movement

B Complete squares and match abbreviations

• Set the completion and the matching task for individual work and pairwork checking.
• Conduct feedback. Ask individual students to spell the words.
• Ask students in pairs to write a sentence using one of the units of measurement from each box.
• Conduct feedback. Ask for examples.

Answers

1
km – kilometre
m – metre
cm – centimetre

in – inch
ft – foot/feet
2
sec – second
min – minute
hr – hour
3
A – ampere
Hz – hertz
kw – kilowatt
W – watt
V – volt
MW – megawatt
4
mg – milligram
g – gram
kg – kilogram
t – ton
lb – pound
5
bar – bar
psi – pounds per square inch
Pa – pascal
kPa – kiloPascal
6
kph – kilometres per hour
mph – miles per hour
fps – feet per second
rpm – revolutions per minute

NB: Students are likely to come across metric measurements, e.g., *metre, kilogram,* etc., (used in Europe) and imperial measurements, e.g., *gallon, foot, pound* (still used sometimes in the US and the UK). This can sometimes cause confusion, e.g., t = the imperial ton (2240 pounds) and T = the metric tonne (1000 kilograms).

C Say pairs of numbers

- Ask students to scan the numbers.

- Choose five different numbers and say them in a random order. Students scan to identify the numbers you have given.

- Read through the numbers with students and elicit what is being measured in each pair. Model the pronunciation. Students repeat for practice.

- Ask students to choose different numbers for their partners to identify. They could then invent their own pairs of numbers to dictate to their partners, who write them down to check.

D Choose greater amounts

- Set the task for pairwork. Do not confirm answers at this point.

Answers

1 b 500 m 2 a 130 mins 3 a 2.9 kg
4 b 3ft 4 ins 5 b 24 kw 6 b 700 mph
7 a 950 mm 8 a 19 bar

E ◉ (CD1 T13) Listen to check answers

- Play the recording for students to listen and check their answers.

- For extra writing, listening and speaking practice, students could write down three more pairs of numbers to test their partners on the greater amounts.

Tapescript
Presenter:
Lesson 5 Talking about units of measurement
E Listen and check your answers.

Voice: 1 Five hundred metres is more than one thousand, five hundred feet.
2 One hundred and thirty minutes is more than two hours.
3 Two point nine kilograms is heavier than four pounds.
4 Three feet four inches is more than seventy-eight point nine five centimetres.
5 Twenty-four kilowatts is more than two hundred and forty watts.
6 Seven hundred miles per hour is faster than one thousand kilometres per hour.
7 Nine hundred and fifty millimetres is longer than nought point one metre.
8 Nineteen bar is greater than nine hundred kiloPascal.

F Discuss road signs

- Refer students to the visuals and ask where they would see signs like these. Point out that Sheffield and Leeds are cities in the North of England. Clarify the words *bridge* and *loading*.

- Set the task as group work.

- Conduct feedback with the whole group. Elicit how these signs are different from the signs in the students' country.

Answers

1 After three quarters of a mile, you mustn't drive over 50 mph.
2 In two miles, there is a bridge. Vehicles over 4.4 m or 14 ft 6 ins can't go under it.
3 The maximum weight allowed is 7.5 T (except for vehicles loading goods).
4 This is a sign on the motorway showing how many miles it is to different places.
5 This sign tells drivers that there is a hill with a 20% gradient.
6 This is a sign for a hospital with an Accident and Emergency department. This department is not open 24 hours a day.
7 This sign tells motorists that after 20 km the road will only be open for people who live there.
8 This is to tell drivers that in 450 m there will be some road works.

CLOSURE

- Elicit other places students might see units of measurement written down, e.g., on the deck of a rig.

Lesson 6: Making calculations

Objective

- to learn terms for and use basic mathematical symbols

Vocabulary

- words related to calculating

LEAD-IN

- Review the previous lesson briefly by eliciting categories of measurements.

- Write two distances on the board and elicit how they are said. Then ask students to add the distances together. Put the answer on the board. Elicit the symbol for adding (+) and the verb *add*. Elicit any other symbols the students know from basic maths and put them on the board.

A Match symbols and names

- Set the matching task for pairwork and feed back orally from different pairs. Explain that these terms are used to describe a calculation. Write a simple one on the board, e.g., *12 + 3 = 15.* Elicit *twelve plus three equals fifteen.*

- Point out that it is important to learn the prepositions that are used with some of the verbs, e.g., *multiply/divide by, subtract from, add to.*

- Give some simple calculations orally for students to work out in their heads. Students can give each other some simple mental calculations in pairs.

Answers

1 plus/add (to)
2 equals
3 minus/subtract (from)
4 is approximately
5 multiplied (by)
6 is not equal (to)
7 divided (by)

B Learn symbols on a calculator

- Write a very complicated calculation for the students on the board, which they are unlikely to be able to do in their heads. Elicit that they need a *calculator* to do the calculation.

- Ask students to try to remember what symbols there are on a calculator without looking at one. They should write them down.

- Set the task for individual work and pairwork checking.

- Conduct feedback. Ask if they had remembered these symbols. Elicit examples of squaring and cubing.

Answers

2 C
3 ^{2}
4 $\sqrt{}$
5 =

C Write calculations

- Give some calculations orally and ask students to write them down using symbols and numbers. Check by asking them to read the calculations back to you. Write them on the board. Students compare what they have written.

- Set the task for individual work and pairwork checking. Point out that calculations would not normally be written in words.

- Conduct feedback. Ask individuals to write the calculations on the board.

- Elicit the answer to calculation 3 by asking *What did you make calculation 3?*

Answers

1   15 x 7.35 = 110.25
2   128 ÷ 12.5 = 10.24
3   12 ÷ 4 x 7 + 11 = 32

D Practise saying calculations

- Ask individual students to read out the calculations. Correct or confirm.
- For further practice, ask students to write a calculation each in numbers. They then take it in turns to read it out. The class write it down and do the calculation. If there is any discrepancy in the answers, write down the original calculation on the board.

E Calculate area

- Draw a shaded rectangle on the board and tell students that you need to buy a new carpet for the office. Elicit that this shaded part is called the area. Ask how we can calculate the area of a rectangle. Ask when they have to calculate area at work.
- Refer students to the *Area* information box in their books and check their answers.
- Set the task for pairwork.
- Conduct feedback. Students answer in their pairs. Do not confirm whether they are right at this point.

F Compare answers

- Students change partners and ask and answer the questions about area. Refer students to the speech bubbles.
- Conduct feedback.

Answers

1   2 cm²
2   162 mm²
3   66 mm²
4   120 mm²

Number 1 has the largest area; number 3 has the smallest area.

CLOSURE

- Discuss with students when and what sort of calculations they have to make at work or in life generally. Elicit whether they use calculators rather than do mental maths if it is a quite simple calculation.

Lesson 7: Measuring dimensions

Objective

• to learn vocabulary for describing measurements and dimensions

Vocabulary

• nouns and adjectives related to dimensions

LEAD-IN

• Ask some simple calculations to review the previous lesson.

• Write some symbols on the board to check students remember the words for them.

A Label dimensions

• Books closed. Use the dimensions of the room you are in to elicit *height*, *width* and *length*. Put these under the heading *Dimensions* on the board. Practise pronunciation as the clusters can prove problematic.

• Books open. Read the *Dimensions* information box with sudents. Elicit what sort of objects they might have to describe using dimensions.

• Set the task for individual work. Conduct feedback.

Answers

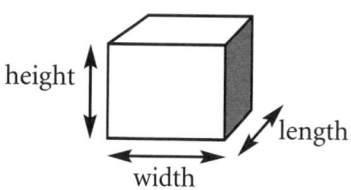

B Categorize adjectives used to describe dimensions

• Elicit adjectives we can use to describe dimensions.

• Refer students to the adjectives in the word box in order to check.

• Set the categorization task for pairwork.

• Conduct feedback.

Answers

Height	Width	Length
tall high shallow deep	wide narrow	short long

C Describe dimensions

• Explain that we can describe dimensions in two ways. Refer students to the information box.

• Give other examples to elicit the two ways of describing height.

• This is probably a good point to introduce *length* and *depth*. Indicate that we can use *shallow* and *deep* to describe water and wells.

• Elicit questions that would give answers about dimensions: *How high/long/wide/deep? What is the height/length/width/depth of ...?* Model and practise the questions for pronunciation, paying attention to the weak form of *of*.

- Set the task for individual work.
- Conduct feedback.

Answers

1 high
2 width
3 length

D Write descriptions giving dimensions

- Ask questions with *how* about places, objects and equipment connected with the oil industry. Point out that when we estimate, we use *about* or *approximately*.
- Refer students to the visuals and set the task for individual work and pairwork checking.
- Conduct feedback. Ask individual students to give some descriptions and compare them. Do not confirm whether they are correct at this point.

E ◑ (CD1 T14) Listen to check answers

- Play the recording for students to listen and compare the descriptions with their own.

Answers

1 The box is 2 centimetres high. Its width is 4.25 centimetres and its length is 6.75 centimetres.
2 The car is 1.4 metres high. It is 1.8 metres wide and 5 metres long.
3 The rig is 65.5 metres high and its area is 35.5 square metres.

Tapescript
Presenter:
Lesson 7 Measuring dimensions
E Listen and compare the descriptions with yours.

Voice: 1 The box is two centimetres high. Its length is six point seven five centimetres and its width is four point two five centimetres.
2 This car is one point four metres high. It is one point eight metres wide and five metres long.
3 The rig is sixty-five point five metres high and its area is thirty-five point five square metres.

CLOSURE

- Review descriptive language by asking students to describe and give the dimensions of things/places where they work.

Lesson 8: Measuring circles and pipelines

Objectives

- to learn vocabulary and practise the language for describing and measuring parts of a circle
- to learn vocabulary and practise the language for calculating circumference, area and volume

Vocabulary

- words related to parts of a circle
- words related to circumference, area and volume

LEAD-IN

- Draw a circle on the board and elicit the words *circle*, *circular* and *round*.

- Ask students to name ten objects that are circular. Ask them to name objects that they work with that are circular.

A Parts of a circle

- Go through the words in the word box.
- Set the task for individual work.
- Conduct feedback.

Answers

2   diameter
3   radius

B Calculate circumference and area

- Elicit how to measure the circumference and area of a circle. Refer students to the *Circumference/Area* information box. Make sure students can say as well as write the formulae.

- Draw a circle on the board and give the radius. Ask students to calculate the circumference and the area of the circle.

- Ask individual students *What did you make the circumference/area?* Elicit answers beginning with *I made it ... .* Repeat with another circle and radius.

Students ask and answer questions to practise the exchange.

- Set the task for individual work and pairwork checking.

- Conduct feedback. Ask students to ask and answer questions in a group to check answers.

Answers

	Circle 1	Circle 2	Circle 3
circumference	3.46 cm	6.28 cm	8.17 cm
area	0.95 cm^2	3.14 cm^2	5.31 cm^2

C Calculate volume

- Draw a piece of pipe on the board and indicate that you want to calculate its volume. Elicit how to do this. Ask students how volume is described.

- Ask students to read the *Volume* information box to check. Label the pipe on the board with the measurements of the pipe in the box. Ask students to explain how to calculate its volume.

- Read the information box to check. Check the meaning of *internal diameter* and *therefore*.

- Refer students to the three pictures. Set the task for individual work and pairwork checking.

- Conduct feedback. Ask individual students to describe the process of measuring the volume for each. They should use the following phrases: *The internal diameter of the pipe is ..., therefore the radius is... . To get the area, I use the formula ..., then I multiply this by the length. The volume is ... .*

Answers

1 9.717 m^3
2 628,200 cm^3
3 11,778.75 m^3

CLOSURE

- Ask students what other objects they might have to calculate the volume of in their work.

- Dictate some of the new words from this lesson: *circumference, radius, diameter, internal, pi, volume* and *barrel*.

Lesson 9: Taking other measurements

Objective

- to study and practise language for measuring and estimating temperature, sound and distance

Language

- language for estimating and guessing

Vocabulary

- words related to measuring sound, temperature and distance

LEAD-IN

- Review measurements that have been covered in the unit so far.
- Ask students to work in pairs to note down any other types of measurements they need to be able to do.
- Conduct feedback. Write any suggestions on the board.

A Do a temperature quiz

- Ask students what the temperature is today and whether they know any other ways of describing temperature.
- Refer them to the *Temperature* information box. Model the pronunciation of the different measuring scales for students to repeat. Check the meaning of *freezing*, *boiling* and *surface*.
- Set the task for pairwork. Divide pairs into Student 1 and Student 2. Students take turns to ask and answer questions to find the answers. Make sure students are looking at the relevant pages.
- Explain that both *freezing* and *boiling* can be used colloquially to say that something is either very hot or very cold: *It's freezing in here!*
- For further practice, ask students to write three more questions about temperature for another pair. Pairs take it in turns to ask and answer questions.

- Conduct feedback. Choose some pairs to ask their questions to the group.

Answers

1 The freezing point of water is 0°C.
2 The boiling point of water is 100°C.
3 The normal temperature of the human body is 37°C.
4 The surface of the Sun is 5,500°C.
5 0°K is −273°C.
6 32°F is 0°C.

B Measure sound

- To elicit how we measure sound, describe a situation when something was so loud it was impossible to hold a conversation. Ask for more examples of situations when the noise level is very high. Elicit what workers have to do when the decibel level is excessively high (wear ear plugs). Ask which jobs require workers to wear ear plugs and whether students have to or have ever had to wear ear plugs.
- Set the task for pairwork. Elicit answers, but do not confirm whether they are correct at this point.

C ◀)) (CD1 T15) Listen to check answers

- Play the recording for students to listen and check their answers. Did any of the answers surprise them?

Answers

1 noisy factory – 100 decibels
2 chain saw – 120 decibels
3 speech at 1 m – 60 decibels
4 moon rocket at 300 m – 200 decibels
5 car horn at 4 m – 80 decibels
6 quiet office – 40 decibels

Tapescript
Presenter:
Lesson 9 Taking other measurements
C **Listen and check your answers.**

Voice 1: Do you have all the answers?
Voice 2: I think so, yes.
Voice 1: Okay, so how loud is a noisy factory?
Voice 2: A fairly noisy factory would be one hundred decibels.
Voice 1: And how loud is a chain saw?
Voice 2: A chain saw is one hundred and twenty decibels.
Voice 1: How loud is speech at one metre?
Voice 2: Generally about sixty decibels.
Voice 1: And how loud is a moon rocket at 300 metres?
Voice 2: A moon rocket – that's very loud. It's two hundred decibels.
Voice 1: So how loud is a car horn at 4 metres?
Voice 2: That's eighty decibels.
Voice 1: And the last one. How loud is a quiet office?
Voice 2: Not very loud – that's got to be forty decibels.

D Measure long distances

- Ask two students in the class where they live and elicit what the distance is between these places. Ask what the question would be (*How far is it from … to …?*). Repeat this with other pairs of locations that students give you.

- Write a distance on the board, e.g., *3 miles.* Say *It is three miles from here to … .* Students guess the destination. It could be your home, a restaurant, a famous landmark, etc. See if they can convert miles into kilometres (or vice versa).

- Read through the *Distance* information box with students. Check they understand the difference between *miles* and *kilometres.* They then ask each other some questions with *How far?*, e.g., *How far is it from your house to your work/the motorway/the bank?* etc.

- Refer students to the table of cities and the map. Set the task for pairwork. Stress that they should guess the distance, but not check their answers yet.

- Elicit estimates from each pair.

E Check answers

- Divide pairs into Student 1 and Student 2. Students take turns to ask and answer questions to find the answers. Make sure students are looking at the relevant pages. They use the information to find out how close their estimates were and to help each other complete the right-hand column of the table.

- Find out whose guesses were most accurate and whether they found any answers surprising.

- Ask how long it might take to drive/fly between the pairs of cities.

Answers

1 London–Moscow: 1,557 miles
2 Prague–Mumbai: 9,334 miles
3 Istanbul–Tehran: 1,270 miles
4 Bangkok–Tokyo: 2,849 miles
5 Kuwait–New Delhi: 1,755 miles
6 Cairo–Singapore: 5,127 miles
7 Rome–Dubai: 2,696 miles
8 Paris–Hong Kong: 8,193 miles

CLOSURE

- Review the lesson by asking students to name the three scales of measuring temperature. See if they can remember the answers to the questions in exercise A.

- Ask what questions they would ask to get the following answers:
  1 It's 75 degrees.
  2 It's 100 decibels.
  3 It's 340 miles.

Lesson 10: Measuring pressure

Objectives

- to learn vocabulary to describe and calculate pressure
- to read about flow and quantities in the context of mud systems

Vocabulary

- words related to measuring pressure
- words related to flow and mud systems

LEAD-IN

- Revise the measurements from the unit by asking students to write down as many measurement abbreviations as they can within a time limit. They can do this in pairs.

- Ask when measuring pressure is important in their job or in the oil and gas industries in general.

A Measure pressure

- Elicit how to calculate pressure. Refer students to the *Pressure* information box to check answers.

- Explain that pressure is measured in different units. Ask students if they know the two types. Refer them to the table. Point out the headings *Imperial* and *ISO*. Elicit any information they may already know about these different measurements.

- Set the table completion task for individual work and pairwork checking.

- Conduct feedback. Ask for answers from individual students.

Answers

	Imperial	ISO
force	pounds	square metres
area	square inches	newtons
pressure	pounds per square inch (psi)	newtons per square metre (pascal/Pa or kilopascal/kPa)

B Convert pressures using formulae

- Ask if any students know how to convert pressures from psi to Pa. If the formula is suggested, put it on the board. Ask how high pressures are measured. Read through the *Conversion formulae* information box with students to check answers.

- Students will need calculators to do the task. If these are available, set the task for individual work and pairwork checking. (Tell them to work out the calculations to a maximum of six decimal places.) If students do not have calculators, suggest that they find the answers by looking at a pressure converter on the Internet at home. Sites such as:
  http://www.convertme.com/en/convert/pressure
  or
  http://www.onlineconversion.com/pressure.htm
  are very useful.

- Conduct feedback. Elicit how students reached their answers.

Answers

1 10 kPa = 10,000 Pa
2 20 psi = 137,900 Pa
3 15 bar = 217.6 psi
4 848085 Pa = 123.004 psi
5 250 psi = 1723.69 kPa
6 1 bar = 14.5038 psi

C Check understanding of vocabulary related to mud systems

- Tell students they are going to read a short text about drilling mud. Elicit what they know about the job of a mud man or engineer and the importance of mud systems when drilling for oil or gas.
- Set the matching task for individual work and pairwork checking.
- Elicit how drilling mud is different to mud that occurs naturally on the ground. Elicit examples of substances that are very viscous and/or dense.

Answers

1 reservoir – A place that stores a large amount of liquid, e.g., an oilfield.
2 mud – A mixture of earth and water.
3 viscosity – How easy it is for a liquid to flow.
4 density – The weight of a unit of a substance.

D Read a text on mud drilling for detailed information

- Before students read the text, tell them to read through the questions.
- Check the meaning of *mixture*, *fluids* and *common*.
- Set the reading task for individual work and pairwork checking.
- Conduct feedback. Ask individual students for answers.

Answers

1 Drill mud controls the pressure of the reservoir fluids.
   It cleans the drill bit and the bottom of the well.
   It carries cuttings out of the well.
2 The mud engineer changes the mixture of the mud to suit the well.
3 No. Barite gives density.
4 Fifty-two sacks.
5 He uses less barite.

CLOSURE

- Revise the meaning and spelling of unfamiliar words from the text. Explain that drilling and mud systems will be covered in more depth in a later unit.

Answers

A

1 b
2 a
3 c
4 c
5 b
6 22,470,000 gallons
7a Saudi Arabia
7b USA
8 millimetre; centimetre; inch; foot; metre; kilometre; mile
9 second; minute; hour; day; week; fortnight; month; season; year
10 6895 Pa; 1 bar; 20 psi; 250 kPa

B

Name of pipeline	Baku-Tbilisi-Ceyhan pipeline (BTC)
Length	1,760 km
Diameter	mostly 42 inches; near Ceyhan 36 inches
Likely capacity in 2009	1 million bbl/ 160,000 m³

 CD1 T16

Tapescript
Presenter:
Unit 2 Review
B Listen to someone describing one of the world's largest oil pipelines. Complete the table with the statistics you hear.

Voice: The Baku-Tbilisi-Ceyhan pipeline (sometimes known as the BTC pipeline) transports crude oil one thousand miles from the Azeri-Chirag-Guneshli oil field in the Caspian Sea to the Mediterranean Sea. The length of the pipeline itself is one thousand, seven hundred and sixty kilometres, so it's one of the longest pipelines in the world. It passes through Azerbaijan, Georgia and Turkey. The pipeline has a forty-two-inch diameter for most of its length, narrowing to a thirty-six-inch diameter when it gets near Ceyhan. Normal capacity, from 2009 onwards, is expected to be one million barrels, in other words, one hundred and sixty thousand cubic metres of oil per day. It has a capacity of ten million barrels of oil. It is hoped that the pipeline will be in use for fifty years.

• Refer students to the self-assessment grids.

Word list

add	fluid
area	foot
avoided	fraction
bar	height
calculate	hertz
centimetre	inch
circumference	injury
clay	kelvin
countable	kilogram
cubed	kilometre
cubic	kilowatt
cuttings	length
decade	liquid
decibel	mass
density	megawatt
diameter	metre
divide	mile
drop	milligram
element	millimetre
equals	minus
fall	multiply
faulty	pascal
flow	percent

DESCRIBING EQUIPMENT

Lesson 1: Talking about workshop tools

Objective

- to identify and describe the uses of a range of hand tools

Language

- phrases to express purpose

Vocabulary

- nouns and verbs related to hand tools

LEAD-IN

- Draw a hammer and a screwdriver on the board to elicit *tools*. Ask where tools can usually be found: *in a workshop*. Ask students if they use hand tools at work or at home and if they know the names of any. Put any suggestions on the board. Do not explain their uses at this point.

A Discuss and label hand tools

- Books open. Refer students to the picture of the workshop with the tools.
- Set the task for pairwork. After talking about the tools, ask students to label the tools they know.
- Conduct feedback. Ask for the names of the different tools. Check spelling. Then find out which tools students have used.
- Ask if students have any of these tools at home.

Answers

(left to right)
Top row: screwdrivers; hammer; spanner; saw
Middle row: spanners; screws; pliers; pipe wrench; nails; calipers; (electric hand) drill
Bottom row: vice; file; chisel; grinder

B Study verbs often used with hand tools

- Refer students to the diagrams and elicit which tools are shown in each.
- Set the task for individual work and pairwork checking.
- Conduct feedback.
- Elicit other contexts when these verbs can be used other than with tools: *tighten/loosen + belt, grip someone's hand, sharpen pencils.*

Answers

1   loosen
2   chip away
3   tighten
4   grip
5   sharpen

C Describe the uses of different hand tools

- Refer students to the lists of tools and uses.
- Set the matching task for individual work and pairwork checking.
- Conduct feedback. Ask different pairs the question: *What do you use a ... for?* Elicit the answer: *You use a ... to ... .*

- Students work in pairs. They use the question to practise the vocabulary.
- Ask if students can suggest any other uses for these hand tools.
- Books closed. Ask students if they can complete the following collocations.

    to tighten
    to grip
    to make a hole in
    to connect
    to rotate
    to hold

Answers

You can use ...
2   a screwdriver to tighten a screw.
3   calipers to measure internal or external dimensions.
4   a file to sharpen other tools.
5   a hammer to connect two pieces of wood with a nail.
6   a saw to cut a piece of wood or metal.
7   a spanner to loosen a bolt.
8   a pipe wrench to rotate a pipe.
9   a drill to make a hole in a piece of wood or metal.
10   a chisel to chip away metal.
11   a vice to hold a piece of wood or metal securely in place.
12   pliers to grip small objects.

D Complete a description

- Tell students that you want to make a bench. Ask what you need and what you have to do.
- Set the completion task for individual work and pairwork checking.
- Elicit answers, but do not confirm whether they are correct at this point.

E 🔊 (CD1 T17) Listen to check answers

- Play the recording for students to listen and check their answers.
- Compare the completed description with what students originally suggested. Ask if they would need to do anything else to make the bench.

Answers

1   cut
2   hold
3   make
4   screw

Tapescript
Presenter:
Unit 3 Describing equipment
Lesson 1 Talking about workshop tools
🅴 **Listen and check your answers.**

Voice: To make a bench, you need a saw to cut the wood, a vice to hold the wood, a drill to make holes in the wood, and a screwdriver to screw the pieces of wood together.

F Give and write descriptions of tools needed for other projects

- Set the discussion task for pairwork. Encourage students to use *You need a ... to ... .* and *You can use a ... to ... .*
- Conduct feedback. Compare suggestions from different pairs.
- For written consolidation, students could think of another two jobs and write a simple description of what is needed. They then take out the verbs to create a gap-fill for their partners.

CLOSURE

- Review the names of tools. Give the uses to elicit the names.
- Elicit the verbs by gesturing.

Lesson 2: Expressing ability

Objective

- to express personal ability and possibility

Language

- modals: *can/can't*

Vocabulary

- words related to equipment and skills

LEAD-IN

- Tell students about someone who can do something extraordinary, e.g., *My daughter is only three, but she can use a computer. My son is ten and he can do a lot of things on a computer that I can't.* Ask for similar stories from students.

A Use *can* and *can't* to express ability

- Draw an example of a Johari window table on the board, similar to the one in the course book. Use the names of two students (A and B) in the group. In the top left-hand box, write a skill that both students can do; in the top right-hand box write something that A can do, but B can't. In the bottom left-hand box, write something that A can't do, but B can; in the bottom right-hand corner, write something that neither A nor B can do.

A can/B can	A can/B can't
A can't/B can	A can't/B can't

- Refer students to the table and set the task for individual work and pairwork checking. Draw attention to the fact that the 3rd-person form has no *s*.

- Conduct feedback. Ask students to give you full sentences. Focus on pronunciation, particularly of *can't* /kɑːnt/ and the weak form of can /kən/.

Answers

1. Bob can *drive a truck*, but Ahmed can't.
2. Bob and Ahmed can both *use a drill*.
3. Bob and Ahmed can't *operate a forklift*.
4. Bob can't *speak Arabic*, but Ahmed can.

B Find and correct grammatical mistakes in a text

- Set the task for individual work and pairwork checking.

- Conduct feedback. Ask individual students for corrections. Do not confirm whether they are correct at this point.

C Write the correct version

- Set the task for individual work. Monitor for errors or problems.

D 🔊 (CD1 T18) Listen to check answers

- Play the recording for students to listen and check their answers and pronunciation.

- If necessary, drill similar sentences for reinforcement.

Answers

My brother's an English teacher and he can read and write English really *well*. I can *speak* English, but I *can't* write it. I'm more practical though. I can use a drill and *repair* things in the house. My brother *can't* do that!

Tapescript
Presenter:
Lesson 2 Expressing ability
D **Listen and check your answers.**

Voice: My brother's an English teacher and he can read and write English really well. I can speak English, but I can't write it. I'm more practical though. I can use a drill and repair things in the house. My brother can't do that.

E **Ask and answer questions using *can/can't* in order to complete a table**

- Check students are familiar with all the vocabulary in the box, e.g., *a fuse, a crane.*
- Set the question and answer task for pairwork. Point out that this is a similar activity to the one at the start of the lesson.
- Do not check answers at this point.

F **Report findings to the class**

- Conduct feedback. Check the results of the task by asking students to give the rest of the class information from their completed table, e.g., *I can ..., but ... can't. We can both ...,* etc.

G **Write sentences about abilities of students in the group**

- Put students into small groups. Students should note down how many people in the group can/can't do different things and complete the sentences in exercise G with this information. They can discuss the skills in the box and/or add ideas of their own, if they wish.

- Conduct feedback. Ask individual students from the different groups to take turns to read out their sentences.

H **Use *can/can't* for possibility**

- Elicit the use of *can/can't* for possibility by giving examples of what you can and can't do with a screwdriver, e.g., *Can you cut wood with a screwdriver?*
- Set the task for individual work and pairwork checking.
- Conduct feedback. Be flexible about the answers, e.g., it could be argued that you can cut electric cable with a pair of scissors, or chip away metal with a screwdriver, but it is not efficient or sensible to do so.

Answers

1   can't
2   can't
3   can't
4   can
5   can't
6   can't
7   can
8   can

CLOSURE

- Review the lesson by asking what students can remember about different students' abilities in the group, e.g., *Ahmed can ..., but Mehdi can't.*

Lesson 3: Describing place and position

Objective

• to describe where things are in a room or workshop

Language

• prepositions of place

Vocabulary

• words related to position and place
• review of names of tools

LEAD-IN

• Establish the concept by asking the positions of certain objects in the room or by turning to page 18 and eliciting where things are in the photo of the rig.

• Ask students where different equipment is kept at their workplace.

A Describe position and place

• Refer students to the picture on page 46. Set the task for individual work and pairwork checking.

• Conduct feedback. See if students can distinguish between the use of *above* and *over*, and *under* and *below*. Although they are often used interchangeably, *above* and *under* are more specific and tend to indicate a direct vertical relationship of objects.

Answers

1   next to
2   above
3   inside
4   below
5   below
6   in front of
7   behind
8   between

B Write sentences using prepositions of place

• Set the task for individual work. Monitor and answer any queries students have.

• Conduct feedback. Ask for example sentences from individual students.

C Describe specific position

• Set the completion task for individual work and pairwork checking.

• Conduct feedback. Ask for answers from individual students.

• Point out the different prepositions *in*, *at* and *on*. Explain that the use of the correct word when giving a description of place and position is very important.

• Refer students to the photo on page 18 or draw a picture on the board with items in different positions to elicit the same language.

• For extra practice, students tell each other about the position of things in a room or on a wall in their house or office.

Answers

1   in the *middle*
2   at the *bottom*
3   at the *front*
4   in the *corner*
5   on the *right-hand side*

D Practise using prepositions and names of tools in an information gap activity

- Explain that students have to ask and answer questions to find eight differences between the two pictures of workshops.

- Set the task for pairwork. Divide pairs into Student 1 and Student 2. Demonstrate by taking the part of Student 1 and choosing a strong student to be Student 2. Ask *Is there a box of nails in your picture?* (*Yes.*) *Is anything written on the front of it?* (*No.*) Then say *In my picture the word Nails is written on the front of the box.* Make sure each student in the pair is looking at a different picture and cannot see their partner's picture. When they have finished, they can look at each other's pictures to check they have noted all the differences.

- Conduct feedback. Ask for differences from different students. Encourage full sentences to practise prepositions and pronunciation.

- For further practice, students can ask and answer questions about the tools in their own workshops, if appropriate.

Answers

In Student 2's picture:

1   The word *Nails* isn't written on the front of the box of nails.
2   The saw above the drill doesn't have a blade
3   There is no plug on the drill.
4   There are two files on the bench.
5   There is only one spanner next to the box of screws.
6   There are two spanners between the hammer and the saw.
7   There are no calipers on the wall.
8   There are only two screwdrivers on the left of the hammer.

E Follow instructions by drawing and labelling a workshop

- Explain what a floor plan is by drawing a floor plan of the room you are in now on the board.

- Check the meaning of *air filter*, *planer* and *router table.*

- Set the task for individual work.

- Ask students to compare their floor plans.

- Conduct feedback. Draw the floor plan on the board and elicit where the various items are.

F Add items to a plan

- Set the task for individual work.

- Students should then explain where the additional items are and their partner should draw them in or point to the place on the diagram where they would be.

CLOSURE

- Write *in*, *at* and *on* on the board and elicit place words that go with these.

- Review prepositions of place by asking where things are in the classroom.

Lesson 4: Describing tools

Objectives

- to read descriptions of parts of tools
- to describe parts of tools

Vocabulary

- verbs connected with tools
- parts of tools

LEAD-IN

- Briefly review the previous lesson by drawing the wall of a workshop on the board and eliciting names of tools from students. Draw the items on the wall. When complete, ask students to give you sentences describing the positions of different tools.

- Ask students to name any power tools they work with.

A Describe an electric drill

- Ask students to tell you what an electric drill is used for and if they know the names of any of its parts.

- Refer them to the diagram and the verbs in the word box. Set the completion task for individual work and pairwork checking.

- Conduct feedback. Do not check answers at this point.

B 🔊 (CD1 T19) Listen to check answers

- Play the recording for students to listen and check their answers.

- Ask if they can name any other tools that have any of the same parts.

Answers

1 is
2 use
3 has
4 supplies
5 holds
6 controls
7 connects

Tapescript

Presenter:

Lesson 4 Describing tools

B Listen and check your answers.

Voice: This is an electric drill. You can use it to make holes in wood, metal or concrete. It has a motor inside it, a trigger, a power cord and plug at the bottom, and a chuck and a bit at the front. The power cord supplies power to the motor. The chuck holds the bit in place. The trigger controls the motor. The plug connects the power cord to the power supply.

C Answer questions about a drill

- Set the task for individual work and pairwork checking. There is no need to insist that students write full sentences.

- Conduct feedback. Ask individual students for their answers.

- Books closed. Ask students to name the parts of an electric drill. Give the verbs *supply, connect, hold* and *control* to elicit what the parts do.

Answers

1   It is at the front.
2   It powers the motor.
3   It is at the end of the power cord. It connects the power cord/the drill to the power supply.

D Study parts of other tools

* Refer students to the two diagrams. Ask if they can give you the names of the tools and say what they are used for, without reading the text.

* Explain that there are mistakes in the descriptions and they should read to find and correct them. Set the task for individual work.

* Do not go through answers at this point.

E ◉ (CD1 T20) Listen to check answers

* Play the first description for students to listen and check their answers.

* Conduct feedback. Ask for the mistakes. Write the new vocabulary on the board and go through pronunciation. Point out that *adjust* is also a verb. Elicit what things you can adjust. Indicate that *jaw* comes from that part of the human body.

* Play the second description and follow the same procedure.

* For additional focus on the texts, elicit words that collocate with *fixed*, *moveable*, *adjusting*, *control*, *safety*, *grinding* and *quenching*.

* If appropriate, students can write a short description of another tool using the texts as models.

Answers

This is a pipe wrench. You can use it to rotate pipes. It has a handle, *one* adjusting *nut*, a fixed jaw and a moveable jaw. The adjusting nut is *behind* the moveable jaw and adjusts the position of the *moveable* jaw.

This is an off-hand grinder. You can use it to recondition tools like screwdrivers and chisels. It has a column, *two* grinding *wheels*, safety screens, rests, a control switch and a quenching tank. The control switch on the *front* of the column operates the grinding wheels. The rests *below* the wheels hold the work in place, and the screens protect the user from debris. The quenching tank *on the front of* the column is used to cool the work.

Tapescript
Presenter:
E Listen and check your answers.

Voice: This is a pipe wrench. You can use it to rotate pipes. It has a handle, one adjusting nut, a fixed jaw and a moveable jaw. The adjusting nut is behind the moveable jaw and adjusts the position of the moveable jaw.

This is an off-hand grinder. You can use it to recondition tools like screwdrivers and chisels. It has a column, two grinding wheels, safety screens, rests, a control switch and a quenching tank. The control switch on the front of the column operates the grinding wheels. The rests below the wheels hold the work in place, and the screens protect the user from debris. The quenching tank on the front of the column is used to cool the work.

CLOSURE

* Review the new vocabulary from the lesson by asking students to name as many parts of the tools as they can.

Lesson 5: Talking about objects

Objective

• to use the passive voice for talking about procedures

Language

• present simple passive

Vocabulary

• nouns and verbs connected with tools and operations

LEAD-IN

• Review the names and parts of different tools.

A Review basic sentence structure

• Write the words *subject*, *verb* and *object* on the board. Elicit a basic sentence from the class that fits the structure, e.g., *I teach English*. Elicit more basic sentences about students in the class, ensuring that there is an object in each.

• Set the task for individual work. Write the sentences on the board.

• Conduct feedback. Ask different students to indicate the subject, verb and object.

Answers

1 A motor (S) powers (V) the drill (O).
2 The chuck (S) holds (V) the bit (O) in place.
3 Drillers (S) operate (V) the machines (O).
4 Ahmed (S) can use (V) an electric drill (O).

B Make passive sentences

• Model the structure by highlighting that in industry, the passive is used to describe procedures more than the active voice, e.g., *Holes are made with a drill* is used more than *We make holes with a drill*.

• Elicit other passive sentences about workshop and industry procedures, e.g., *How are tools sharpened? How is oil transported?*

• Write example sentences on the board. Highlight the form and the use of the verb *be*. Point out that if we want to include the person or thing that did the action, we call this *an agent*. It is added to the end of the sentence with *by*.

• Read through the information box with students.

• Set the task for individual work.

• Conduct feedback. Point out that *by Ahmed* would not be included in sentence 4.

Answers

2 *The bit is* held (in place) by the chuck.
3 The machines are operated *by drillers*.
4 An electric drill can be *used*.

C Form passive sentences

• Ask students to look at the passive sentences from exercise B and complete the rule in the information box.

• Elicit examples of situations where the passive voice would normally be used, e.g., in formal documents and contracts, in instructions for procedures and rules, in descriptions of processes and in incident reports.

Answers

The passive voice is formed using the auxiliary verb *be* and the *past participle*. It is not always necessary to include *by + agent* in a passive sentence. About 25% of sentences in academic texts use the *passive* voice.

D Identify passive sentences

- Elicit whether students use, or have used, a grinder. Check the meaning of *polish*, *attach to* and *goggles*. Set the task for individual work and pairwork checking.
- Conduct feedback with the whole group.
- Elicit the active forms of the sentences, starting with *we* as the subject when there is no other given.

Answers

are used; can be used; are attached; are held; should be worn

E Write past participles

- Elicit which part of the verb is called the past participle (the 3rd-person form). Point out that in all regular, and in some irregular verbs, it is the same as the past simple (2nd-person form), e.g., *hold, held, held*. However, in some irregular verbs it is different, e.g., *wear, wore, worn*.
- Set the task for individual work, then have brief class feedback. Check students know the past simple of these verbs too.

Answers

1 used 2 monitored 3 measured 4 made
5 shown 6 held 7 worn 8 attached
9 turned 10 taken 11 connected
12 drilled

F 🔊 (CD1 T21) Make active sentences passive

- Introduce and check the meanings of *the flow*, *a pointer* and *a float*.
- Set the task for individual work and pairwork checking.

- Play the recording for students to listen and check their answers.

Answers

1 is connected
2 is used
3 are made
4 is monitored
5 is shown by a pointer
6 be held in place with a pipe wrench
7 be measured with a float
8 cannot be turned clockwise

Tapescript
Presenter:
Lesson 5 Talking about objects
F Listen and check your answers.

Voice: 1 The drill is connected to the energy supply by the plug.
  2 A file is used by the engineer to finish the metal.
  3 The holes are made with an automatic drill.
  4 The flow is monitored by a computer.
  5 The change in level is shown by a pointer.
  6 Pipes can be held in place with a pipe wrench.
  7 Liquid in a tank can be measured with a float.
  8 The screw cannot be turned clockwise.

CLOSURE

- List some different workers in the oil industry, e.g., *the drilling crew*, *the seismic crew*, *the crane operator*, or workers from different office departments, e.g., *the IT Department*. Ask students to think of equipment that is used by each group or person and make sentences, e.g., *Geophones are used by the seismic crew*.

Lesson 6: Describing measuring devices (1): pressure and temperature

Objective

- to read about and learn vocabulary for measuring devices and parts of measuring devices

Language

- language for describing equipment and processes

Vocabulary

- words related to measuring devices

LEAD-IN

- Review passive sentences and vocabulary from the previous lesson by writing some appropriate active sentences on the board, then eliciting the passive forms. Use tool functions from the lesson.

A Define parts of measuring devices

- Elicit the names of common devices used for measuring, e.g., *speedometer* and *thermometer*. Write the words on the board with the things they measure.
- Draw a diagram of a simple thermometer and ask if students can name any of the parts.
- Set the matching task for individual work.
- Conduct feedback with the whole group.
- Ask if students can now name any of the parts of the thermometer on the board. Elicit where else we can see these items.

Answers

1 bulb – The rounded part of an instrument or vessel.
2 pointer – A thin piece of metal that points at a scale.
3 tube – A cylinder made of glass, metal, cardboard or plastic.
4 spiral – A curve that turns around a central point in the shape of waves.
5 scale – A series of numbers or marks at set intervals.

B Read types of pressure-measuring equipment

- Elicit what instruments are used to measure pressure.
- Refer students to the descriptions and visuals. Check or clarify the meaning of *corrugated* and *metal-walled*.
- Set the matching task for individual work and pairwork checking.
- Conduct feedback. Practise the pronunciation of the names of the devices. Ask if any students know when and where the different types of equipment are used.

Answers

1 b
2 a
3 d
4 c

C Label parts of a filled system thermometer

- Write the word *thermometer* on the board and elicit what it measures. Ask for the names of different types of thermometers and whether students use any in their work.

- Refer students to the diagram and the words in the word box. If necessary, model the words for students to repeat, then set the labelling task for individual work. Encourage students to use the glossary at the back of the course book to check the meaning of the technical terms, i.e, *bourdon tube*, *capillary tube* and *bellows*.

- Feed back by asking for answers from the whole group.

Answers

1   bourdon tube
2   capillary tube
3   bulb
4   pointer
5   temperature scale

D Match sentence halves to complete the description of how a device works

- Ask students if they know how the filled system thermometer works. Elicit suggestions.

- Check the meanings of *mercury*, *uncurl* and *indicate*.

- Explain that a short text about how a filled system thermometer works has been broken down into sentence halves.

- Set the matching task for individual work and pairwork checking.

- Conduct feedback. Ask individual students for answers. Do not confirm whether they are correct at this point.

E 🔊 (CD1 T22) Listen to check answers

- Play the recording for students to listen and check their answers.

- Point out how the intonation rises and falls when giving a list, e.g., *a bulb*, *a capillary tube*, *a bourdon tube*, *a pointer and a scale*. Elicit other examples of short lists for students to practise the intonation pattern.

- Highlight useful language for structuring a description, e.g., *This is a ... . It consists of ... .*

Answers

This is a filled system thermometer. It consists of a bulb, a capillary tube, a bourdon tube, a pointer and a scale. This type of system is completely filled with a liquid, usually mercury. When the mercury in the bulb expands, it goes through the capillary tube and into the bourdon spiral. The spiral uncurls, and this movement makes the pointer move. The pointer then indicates the temperature on the scale.

Tapescript
Presenter:
Lesson 6 Describing measuring devices (1): pressure and temperature
E Listen and check your answers.

Voice: This is a filled system thermometer. It consists of a bulb, a capillary tube, a bourdon tube, a pointer and a scale. This type of system is completely filled with a liquid, usually mercury. When the mercury in the bulb expands, it goes through the capillary tube and into the bourdon spiral. The spiral uncurls, and this movement makes the pointer move. The pointer then indicates the temperature on the scale.

CLOSURE

- Review new vocabulary from the lesson.

Lesson 7: Describing measuring devices (2): level

Objectives

- to listen to descriptions
- to describe how level-measuring devices work

Language

- present simple passive

Vocabulary

- words related to level-measuring devices

LEAD-IN

- Briefly review vocabulary and language from the previous lesson by asking students to give you a short description of how a filled system thermometer works.

A Label a diagram

- Refer students to the diagram and the words in the word box. Give them time to look at it and decide what the diagram shows before setting the labelling task for individual work.

B Discuss the measuring device

- Ask students to check in pairs.
- They should then discuss what they think the device measures and describe the movements involved.
- Ask for suggested answers, but do not confirm whether they are correct at this point.

C ◀))) (CD1 T23) Listen to check answers

- Play the recording for students to listen and check their answers.
- Ask what the device is and what it measures. Ask students where devices like this are used. If necessary, students can check the meaning of *tank* and *float* in the glossary.

- Elicit what verbs they heard on the recording. Write the verbs on the board (bare infinitives): *show, use, measure, connect, indicate, decrease, get lower, pull, increase* and *get higher*. Check meanings.

Answers

1 scale 2 pointer 3 tank 4 float

It's a float system and it measures the level of a liquid in a tank.

Tapescript
Presenter:
Lesson 7 Describing measuring devices (2): level
C Listen to the description of the device.

Voice: One method to measure the level of a liquid in a tank is with a float system. A wire connects a float to a counterweight through a system of pulleys. A pointer on the counterweight indicates the level on a scale. As the level of the liquid decreases, the float gets lower and the counterweight is pulled higher. As the level of the liquid increases, the float gets higher and the counterweight is lowered. The change in level is shown by the pointer.

D Complete a description using verbs in the active or passive form

- Give students some present simple active sentences, e.g., *The pointer indicates the level.* See if they can convert them into passive constructions. Reinforce the fact that the auxiliary *be* (*is/are* in this case) must always precede the main verb in a passive construction.

- Set the task for individual work and pairwork checking.

- Ask students to practise reading the full text to each other, then ask for one or two students to read the text to the class as feedback.

- Ask students to close their books and select students to explain the diagram from memory.

Answers

1 is connected
2 indicates
3 decreases
4 gets
5 is pulled
6 increases
7 gets
8 gets
9 is shown

E Discuss how a flow level meter works

- Elicit the names of other devices used to measure level that the students know or work with.

- Refer students to the diagram and elicit or explain that this meter measures flow level.

- Refer students to the words in the word box in exercise F. Ask a few questions to help focus students, e.g., *What do you think rotates? What substance is likely to flow? What rises and falls? What do you think the top of the float indicates?* Set the discussion task for pairwork.

- Conduct feedback. Ask different pairs to describe how they think it works. Note on the board any verbs they use in their descriptions.

F Write a description of a meter

- Check the verbs in the word box against the verbs on the board.

- Read the introductory sentence and check meaning of *consists of*. Reinforce this by asking what the float system on page 58 consists of.

- Set the writing task for individual work. If necessary, provide some of the vocabulary or phrases from the example answer below. Monitor to advise.

- When students have finished, ask them to read their partners' descriptions and compare what they have written.

- Ask one or two students to read out their descriptions to the class.

Example answer

A rotameter consists of a glass tube with a measuring scale on it. It measures flow rate. The float inside it rotates as it rises and falls. The top of the float indicates the rate of the flow of liquid or gas.

CLOSURE

- Ask students to choose words from the lesson to dictate to their partners for a quick spelling test.

Lesson 8: Describing how tools work

Objectives

- to read and listen to descriptions of workshop tools and identify parts
- to complete and write descriptions of workshop tools

Language

- the passive voice with the present simple and modal *can*

Vocabulary

- parts of tools
- words related to position and movement

LEAD-IN

- Draw some simple pictures of tools on the board, e.g., an electric drill, a pipe.
- Elicit parts of the tools and where they are in relation to each other to revise prepositions.

A Describe how a bench vice works

- Write *bench vice* on the board and ask students what it is used for. Elicit any parts of a bench vice that they already know in English and put them on the board, e.g., *jaw*, *screw*.
- Books open. Refer students to the diagram of a bench vice and the words in the word box. Ask for any possible answers to complete the labelling, but do not confirm at this point.
- Set the reading task for individual work and pairwork checking. Students should be able to guess what the *slide*, *body* and *fixed jaw* are, even if they do not know the words, from the description and prepositions.
- Conduct feedback with the whole group. Ask *How is it connected to the bench?* (*by bolts*) *What does it consist of?* (*a main body, a fixed jaw, a moving jaw, a slide, a handle*) *What happens when the handle is turned clockwise?* (*the slide moves towards the bench and*

holds the work)/*anti-clockwise?* (*the slide moves away from the bench and releases the work*)

Answers

1 fixed jaw 2 screw 3 body 4 slide 5 bolt

B ◆ CD1 T24 Label a bench drill and describe how it works

- Elicit what a bench drill is and what it is used for. Write *bench drill* on the board and ask students what it is used for. Elicit any parts of a bench drill that they already know in English and put them on the board, e.g., *controls*, *base*.
- Books open. Refer students to the diagram of a bench drill and the words in the word box. Ask for any possible answers to complete the labelling, but do not confirm at this point.
- Play the recording for students to listen and label the diagram.
- When they have checked answers with a partner, conduct feedback. Explain the basic meaning of *to lock*. Ask what happens when the operating lever is turned clockwise/anti-clockwise to elicit the verbs *raise* and *lower*.
- Revise positions and names of parts by asking *Where is the drilling table? Where is the locking handle? Where are the controls?* etc.

Answers

1 chuck 2 controls 3 motor housing
4 operating lever 5 locking handle 6 base

Tapescript
Presenter:

Lesson 8 Describing how tools work

B **Listen to the description of how a bench drill works, then finish labelling the diagram with the words in the box.**

Voice: This is a bench drill. It can be used to drill holes in wood or metal. It has a drilling table, an operating lever, a chuck, a guard, a locking handle, a motor and controls. The base of the drill is connected to the bench by four bolts. Above this is the drilling table and the locking handle: this is the small handle behind the drilling table. Above this, at the top of the drill, are the controls. They are at the back of the drill, behind the motor housing. The operating lever is on the side of the motor housing and when it is turned clockwise, the drill is lowered. When it is turned anti-clockwise, the drill is raised. The chuck holds the drill bit in place above the drilling table. There is a guard in front of it to protect the user.

C Complete a description of a vernier caliper

- Refer students to the diagram of a vernier caliper. Elicit what they know about calipers, their parts and their use in the petroleum industry. Check the meaning of *clamp* and *exact*.

- Refer students to the gapped text and the phrases in the word box. Set the completion task for individual work and pairwork checking. If they have difficulties, refer them to the positioning of the same phrases in the description of the bench vice in exercise A.

- Conduct feedback.

Answers

1 This is a 2 It can be used to 3 It consists of
4 is used to 5 is used to 6 hold 7 is used to

D Write a description of a hacksaw

- Refer students to the diagram. Read through the names of the different parts.

- Set the writing task for individual work.

E 🔊 (CD1 T25) Listen to compare descriptions

- Play the recording for students to listen and compare their descriptions. They do not have to be the same. Ask students to add any extra information.

- Ask some students to read their descriptions to the class.

Answers

1   hacksaw
2   cut metal
3   a blade, an adjustable frame, a handle, a blade-tensioning screw and a frame-locking screw

Tapescript
Presenter:

E **Listen and compare your description.**

Voice: This is a hacksaw. It can be used to cut metal. A hacksaw consists of a blade, an adjustable frame and a handle. There is also a blade-tensioning screw joining the blade to the frame. It is turned to make the blade tighter. Another screw connects the frame to the handle and keeps it rigid. This is called the frame-locking screw.

F Ask and answer questions about how a tool works

- Demonstrate the task by choosing a tool yourself and eliciting questions about it from students. Set the task for pairwork.

- Conduct feedback by asking the same questions to the whole class.

CLOSURE

- Test new vocabulary. Give definitions and elicit names. Say *They are at the back of the drill. You use these to attach a tool to the bench.*

Lesson 9: Describing types of pumps

Objectives

- to become more familiar with reading technical descriptions
- to learn language for defining types of pumps

Vocabulary

- words related to types and parts of pumps

LEAD-IN

- Write the word *pump* on the board and ask what a pump does.

A Define a pump

- Books open. Set the task for pairwork. Read the definitions aloud as the students read them in their books.
- Ask for answers and record how many 1s, 2s or 3s on the board. Ask students to justify why they made a particular choice.
- Explain or elicit that definition 3 is the best answer because pumps do not only move liquids, they move gas and air too.

Suggested answer

3

B Discuss and list types of pump

- Put students in small groups and ask them to talk about and note down all the types of pump we can see or use in everyday life, e.g., washing machine pumps, hydraulic pumps, etc. This could be done as a competition.
- Conduct feedback and write a list on the board.
- Ask groups to name different pumps used in their work. Conduct feedback and list the pumps on the board as before.

C Match pumps and usage

- Explain that the pumps described in this exercise are common types of pump found on site. Read through the list with students and check if they are in the list on the board.
- Set the task for individual work and pairwork checking.
- Check the meaning of *condenser*, *boiler* (*to boil*), *treater*, *turbine* and *compressor*.
- Conduct feedback.

Answers

1 main and auxiliary oil pumps – Used to move the oil through the pipeline as required.
2 fuel oil pump – Used in oil-fired treaters to pump fuel oil to the burners.
3 lubricating oil pump – Used to circulate oil around a machine such as a turbine, engine, pump or compressor.
4 circulating water pump – Also called a cooling water pump. It is used to pump water through a heat exchanger such as a condenser or oil cooler.
5 chemical feed pump – Small capacity units are used to pump chemicals into boilers; larger units are used as process pumps.
6 fire pump – Used to supply water to plant fire lines.

D Identify distances involved in pumping activity

- Refer students to the diagram and elicit the names of the different items shown, e.g., *water*, *pump*, *supply tank*, *discharge tank*, *pipe*. Clarify and check any problem vocabulary. (Students may wish to label these.)

- Set the task for individual work and pairwork checking.

- Write on the board *suction*, *discharge*, *intake*, *surface*, *source of supply*, *sum*, *thus*. Ask students to work in pairs to deduce the meanings of the words they are unfamiliar with from the context and by looking at the diagram.

- Conduct feedback. Ask in what other situations these words could be used.

Answers

1 Static suction lift C
2 Static discharge head B
3 Total static head A

CLOSURE

- Ask students to refer back to the pumps they listed in exercise B. Can they explain how any of these work?

Lesson 10: Describing how pumps work

Objectives

- to review and extend vocabulary for describing pump systems
- to review and contrast passive and active voices

Language

- active and passive voices

Vocabulary

- words related to pumps

LEAD-IN

- Review the previous lesson by describing the uses of the different pumps to elicit the names. Ask students to describe and name the distances that are important in pump systems.

- Pre-teach the words *valve*, *piston* and *plunger*. Explain that *piston* and *plunger* mean the same.

A Discuss and label a pump system

- Direct students to the diagram. Elicit or explain that it shows a modern domestic heating system pump. Ask students to name different parts of this system.

- Set the task for pairwork and write the following labels: *storage tank*, *pipe*, *valve*, *valve box*, *waste water*, *suction head*, *total static head*, *drive pipe* and *waste valve* on the board. Students should try to label as much as they can.

- Put two pairs together to compare their answers.

- Conduct feedback. If possible, put this diagram on an OHP or draw it on the board.

Answers

1 storage tank 2 pipe 3 total static head
4 discharge valve 5 waste water
6 supply head 7 drive pipe 8 valve box
9 waste valve

B Choose the correct verbs to complete sentences

- Set the task for individual work and pairwork checking.

- Conduct feedback by eliciting differences between the two verbs from individual students. Ask for other examples of equipment or items that can be used with these verbs.

Answers

1 supply (*supply* means 'provide', 'give to', whereas *discharge* means 'come out of')
2 circulates (*circulates* means 'go through or around the system', whereas *reverses* means 'go backwards')
3 drawn (*drawn* means the liquid is 'sucked or pulled' into the pump, whereas *injected* means it is 'put or pushed' into it)
4 moves (*moves* implies 'two-way direction', whereas *lifts* is only 'one way' (up))
5 forces (*forces* implies 'pushing away from', whereas *pulls* implies 'pulling towards')

C Describe different pumps

- Explain that the students are going to complete texts about two types of pump. Refer them to the diagram at the bottom of page 64 and elicit where the water is drawn in (suction valve), where it is forced out (discharge valve) and what makes this happen. Discuss the form of the valve, in this case two balls (ball checks).

- Write the names of the two pumps on the board: *reciprocating* and *single-acting*. Remind students that the passive form is often used in written, more formal descriptions. Set the first completion task for individual work and pairwork checking.

- Conduct feedback. If the diagram hasn't been previously identified, ask if this text describes it or not, and why.

- Elicit the meaning of *to-and-fro*.

- Set the next completion task for individual work and pairwork checking. If students are not sure of the answers, reassure them that they can listen and write them in afterwards.

D 🔊 (CD1 T26) **Listen to check answers**

- Play the recording for students to listen and check their answers and fill in any gaps.

- Go through the text again and ask students to give you the correct answers and say if it is active or passive.

- Books covered. Elicit which verbs are used in the two descriptions and put them on the board. Then ask comprehension questions to elicit answers, using the verbs in the correct form: *What happens when the plunger moves from right to left? What happens when the plunger reverses? What does the pressure of the liquid do? What do we call the movement of the plunger in one direction?*

Answers

1 is produced 2 is drawn 3 (it is) forced

- Explain that it is not always necessary to repeat the full passive form when a sequence is described.

Answers

1 moves 2 is drawn 3 reverses
4 moves 5 is forced 6 is forced 7 is forced
8 is called 9 moves 10 is called 11 takes
12 is discharged 13 is called

Tapescript
Presenter:
Lesson 10 Describing how pumps work
D **Listen and check your answers.**

Voice: Reciprocating pump
In this type of pump, the pumping action is produced by the to-and-fro (reciprocating) movement of a piston or plunger within a cylinder. The liquid is drawn into the cylinder through one or more suction valves, then forced out through one or more discharge valves by direct contact with the piston or plunger.

Single-acting pump
When the plunger moves from right to left, the liquid is drawn into the cylinder through the suction ball check. When the plunger reverses and moves from left to right, the liquid is forced out through the discharge ball check. The discharge ball check is forced open by the pressure of the liquid and, at the same time, the suction ball check is forced closed. The movement of the plunger in the cylinder in one direction is called the stroke of the plunger. The distance the plunger moves in and out of the cylinder is called the length of the stroke. Only one side of the plunger takes part in the pumping action, and water is discharged only during one out of every two strokes. For these reasons, the pump is called 'single-acting'.

CLOSURE

- Refer students to the diagram of the pump again and ask them to explain how it works to their partner, without looking at the descriptions.

Answers

A

Student A's picture:

1   The drill has not got a *bit* in it.
2   The box of bits is lying *next to* not *behind* the drill.
3   The caliper jaws are *closed*, not *open*.
4   There is no *wrench*.
5   The bench vice is holding a piece of *metal*, not *wood*.
6   There is no *hacksaw*.

B

1   **Valve A:**   Butterfly valve.
    Valve B: Swing check valve.
2   **Valve A:**   Yes.
    Valve B: No.
3   **Valve A:**   A circular flap or disc.
    Valve B: A disc attached to a hinged arm.
4   **Valve A:**   When the disc is parallel to the flow – operated by the handle.
    Valve B: When pressure of liquid holds the disc open.
5   **Valve A:**   When the disc is at a right-angle to the flow – operated by the handle.
    Valve B: When the flow stops or pressure downstream of the valve is higher than upstream, the hinged arm swings down and the disc covers the passage.

C   **D**

Students' own answers.

Assess your skills: Describing equipment

• Refer students to the self-assessment grids.

Word list

anti-clockwise	jaw
bellow	lift
bench	loosen
bit	lower
blade	oil field
bolt	piston
bourdon	pliers
bulb	plunger
calipers	pointer
capillary tube	power
chip	practical
chisel	reciprocating
chuck	rotate
circulate	saw
clamp	scale
clockwise	screw
connect	screwdriver
corrugated	sharpen
counterweight	slacken
device	slide
discharge	spanner
draw	spiral
file	stroke
finish	suction
float	supply
flow	switch
force	tank
goggles	tighten
grinder	trigger
guard	tube
hack saw	valve
hammer	vice
handle	wire
hold	wrench
injury	

GIVING INSTRUCTIONS AND WARNINGS

Lesson 1: Following instructions

Objectives

- to practise language for giving and sequencing instructions
- to describe ongoing situations and activities

Language

- the imperative

Resources

- If possible, bring enough paper clips to the lesson so that you can give one to each student.

LEAD-IN

- Establish the concept of giving instructions by telling students to do a sequence of actions such as *Open your books. Turn to page 34. Take a piece of paper. Write down the first word on the page. Tell me the word.* Then ask them to repeat your instructions, word for word, if they can. Elicit that these are *instructions*. Ask for other typical instructions you give the class during a lesson.

- Ask how to make the first instruction negative: *Don't open your books.*

A Match instructions and pictures

- Set the matching task for individual work.
- Go through the answers. Check the meaning of *drop*. Contrast with *fall*.

Answers

1 b 2 c 3 a

B Study the use of the imperative

- Write one of the instructions on the board and elicit that we call the verb form *the imperative*. Ask students to read

the information box and elicit the missing word.

- Explain that we use the imperative to give instructions (as they have seen), warnings and commands.

Answer

don't

C Reorder words to make instructions

- Remind students of the meaning of *horizontal*, *vertical*, *divide* and *section*.

- Tell students they are going to reorder words to find the instructions for making something with a piece of paper and a paper clip.

- Set the task for pairwork. Remind students that the first word of each instruction will start with a capital letter.

D 🔊 (CD1 T27) Listen to check answers and follow instructions

- Play the recording for students to listen and check their ordering of the instructions.

- Play the recording again for students to follow the instructions, pause where necessary. See what they make. What is it?

- Ask if students know any other paper-folding activities. If they do, ask them to instruct each other in pairs or small groups and then instruct you.

Answers

1 Cut a piece of paper three inches by four inches.
2 Draw a horizontal line across the paper one inch from the top.
3 Draw two vertical lines to divide the bottom of the paper into three equal parts.
4 Carefully tear along the vertical lines up to the horizontal line.
5 Fold the left section of the paper up towards you.
6 Fold the right section of the paper up away from you.
7 Put the paper clip on the bottom of the middle section.
8 Hold the paper by the top above your head and drop it!

Tapescript
Presenter:
Unit 4 Giving instructions and warnings
Lesson 1 Following instructions
D **Listen and follow the instructions.**
What do you make?

Voice: 1 Cut a piece of paper three inches by four inches.
2 Draw a horizontal line across the paper one inch from the top.
3 Draw two vertical lines to divide the bottom of the paper into three equal parts.
4 Carefully tear along the vertical lines up to the horizontal line.
5 Fold the left section of the paper up towards you.
6 Fold the right section of the paper up away from you.
7 Put the paper clip on the bottom of the middle section.
8 Hold the paper by the top above your head and drop it!

E Complete a set of instructions

- Ask students how often they send letters these days, and when. Do they prefer to send e-mails or texts? Why/Why not?
- Set the task for individual work and pairwork checking.
- Feed back answers from pairs by asking them to give you the full sentences.
- Practise pronunciation by getting students to read a sentence each in pairs.
- Books covered. Elicit the sequence. Then elicit which words were used at the beginning of the instructions. Write them on the board.
- Books uncovered. Students check the order and the sequence words.

Answers

1 write 2 fold 3 put 4 Write 5 put/stick
6 take; post

F Give and write instructions

- Elicit examples of other simple things to give instructions on, e.g., how to send an e-mail, how to send a text message, how to make a cup of tea/coffee, how to record a film from TV, etc.
- In pairs, students should choose a different activity each to instruct the other on. Preteach or check some useful verbs before they start, e.g., *key in*, *switch on*, *type in*, *click on*, *press*, etc.
- Students complete the task orally.
- Conduct feedback by asking some students to give their instructions to the group. The rest of the class have to guess as quickly as possible what the instructions are for.
- Ask students to write the instructions in their books.

CLOSURE

- Ask students what instructions you usually give them at the end of a lesson. Get them to do them!

Lesson 2: Describing controls

Objective

- to practise language for giving and following instructions for using controls

Vocabulary

- verb-noun collocations

LEAD-IN

- Discuss what sort of instructions students have to give at work, and to whom. Review instruction-giving by asking students for examples of instructions they have given so far today.

- Elicit a negative imperative from the previous lesson.

A Describe controls

- Elicit controls that are found on tools and equipment, e.g., *handles*, *levers* and *buttons*. Write them on the board. Refer students to the visuals and elicit what controls they can see. Pictures 2 and 5 can both be described as buttons or switches.

- Check the words in the word box. Set the completion task for individual work and pairwork checking. Emphasize the fact that more than one verb is sometimes possible.

- Conduct feedback.

Answers

2 *Push/Press* the button.
3 *Push/Pull* the lever.
4 *Turn* the handle.
5 *Press/Push/Turn on/off* the switch.
6 *Pull/Release* the trigger

B Discuss visuals

- Elicit a few examples of where these particular controls might be found.

- Students discuss what each control is and what sort of equipment has them.

- Discuss with students which ones they use regularly, occasionally or never.

C Practise verbs used in describing controls

- Set the completion task for individual work and pairwork checking.

- Conduct feedback. Read the sentences and ask students to supply the verbs. Write them on the board. Do not confirm answers at this point.

D ◉ (CD1 T28) Listen to check answers

- Play the recording for students to listen and check their answers.

- Play the recording again for students to repeat the sentences for practice. Highlight the words that are stressed in the two halves of the sentences (see the highlighted words in the Answers). Give more practice by reading the first half of the sentences yourself and getting students to finish them. Then ask students to finish each other's sentences in pairs.

Answers

1 *Push* the lever *up* to turn the machine *on*, and *push* it *down* to turn the machine *off*.
2 *Pull* the trigger to start the drill, and *release* the trigger to *stop* the drill.
3 Turn the *dial* **clockwise** to increase the flow rate, and *turn* it *anti-clockwise* to *decrease* the flow rate.
4 *Press* the button **on the** *top* to start the machine, and *press* the button **on the bottom** to *stop* the machine.
5 Press the **top** *button* to move up, and *press* the *bottom* button to move *down*.
6 *Turn* the handle *clockwise* to *open* the valve, and *turn* it *anti-clockwise* to *close* the valve.

Tapescript
Presenter:
Lesson 2 Describing controls
D **Listen and check your answers.**

Voice: 1 Push the lever up to turn the machine on, and push it down to turn the machine off.
2 Pull the trigger to start the drill, and release the trigger to stop the drill.
3 Turn the dial clockwise to increase the flow rate, and turn it anti-clockwise to decrease the flow rate.
4 Press the button on the top to start the machine, and press the button on the bottom to stop the machine.
5 Press the top button to move up, and press the bottom button to move down.
6 Turn the handle clockwise to open the valve, and turn it anti-clockwise to close the valve.

E Complete the instructions on using a car

- Lead into the task by miming or eliciting some of the actions that people do when they operate a car, e.g., braking, turning the steering wheel, unlocking it, etc.
- Set the task for pairwork discussion before students write their ideas individually.
- Conduct feedback. Elicit different suggestions.

Example answers

1 *Turn the key clockwise/Press the button/switch* to lock the car.
2 *Turn the key anti-clockwise/Press/Release the button/switch* to unlock the car door.
3 *Turn* the steering wheel *anti-clockwise* to turn left.
4 *Press/Push* the foot brake *down* to stop the car.
5 *Press/Push* the accelerator *down* to *increase* your speed.

CLOSURE

- Elicit instructions on how to operate other machinery that students are familiar with in the workplace.

Lesson 3: Giving instructions for using tools

Objective

- to practise using instruction language for the workshop/using tools

Language

- the imperative
- using sequencers

Vocabulary

- verbs for giving instructions
- sequencers

LEAD-IN

- Review the previous lesson by asking which verbs we use with certain words, e.g., *handle*, *lever*, *dial*, *switch* and *button*. Elicit which of these actions the students have performed today, e.g., pressed the button in the lift, turned the dial on the microwave, etc.

A ◀)) (CD1 T29) Give instructions to drill a hole

- Tell students you need to drill a few holes in the wall. Elicit what tool(s) you need, then ask for some instructions on how to drill a hole. Insist on detailed instructions.

- Books open. Ask students to read through the list of instructions and check the meanings of *mark* and *attach*.

- Set the task for individual work. Students listen and number the instructions, then check in pairs.

- Play the recording again and pause it after each instruction to check. Ask students for the opposites of these verbs used in the conversation: *tighten*, *attach*, *connect* and *start*. Point out the adjectives *tight* and *loose*. Discuss in what other situations the verbs might be used.

- Elicit which sequencing words were used in the conversation. Write them on the board. Leave the information on the board as this will be useful for the final activity.

Answers

1 B 2 D 3 J 4 G 5 I 6 H 7 K 8 E 9 F
10 M 11 A 12 L 13 C

Tapescript
Presenter:
Lesson 3 Giving instructions for using tools
A Bob wants to drill a hole. He asks Vasily for instructions. Listen and put the instructions in the right order, A to M.

Bob: How do I drill a hole in this wall?
Vasily: Well, you need to use a drill. First, measure the work and then mark the hole.
Bob: No problem. What next?
Vasily: Okay, next attach the drill bit and then tighten the chuck.
Bob: Yep, and after that?
Vasily: After that, connect the drill to the power and place the bit over the mark. When you're ready, start the drill and drill the hole.
Bob: What do I do when I finish the hole?

Vasily: First, you stop the drill and secondly, remove the bit from the hole and then disconnect the drill from the power. Finally, you loosen the chuck and remove the drill bit.

Bob: Thanks a lot, that's a great help.

Vasily: My pleasure.

B Give instructions for dos and don'ts in a workshop

- Set the task for pairwork.

- Conduct feedback by asking which instructions students have ticked, and why. They should also say why they did not tick the other instructions.

- Check the meanings of *avoid*, *adjust* and *take off*.

- Explain what instructions you have to follow in your workplace, then ask students to tell each other what instructions people should follow in their own workplace. This could be on site or in an office.

Answers

2

3

5

7

C Give instructions

- Set the task for pairwork. Divide pairs into Student 1 and Student 2. Make sure students are looking at the relevant pages.

- Ask students to make a real conversation by using questions such as Bob did in exercise A. They should also try to use sequencing words when giving their instructions.

- Monitor and advise where necessary. Students check instructions by looking at the relevant pages.

- If appropriate, ask some pairs to repeat their instruction conversations to the class.

Answers

Changing a light bulb
1 Check that the power is off.
2 Take out the old bulb and dispose of it carefully.
3 Screw in the new bulb.
4 Turn on the power and check that the light works.

Cutting a piece of pipe
1 Place the piece of pipe in a vice.
2 Adjust the vice to secure the pipe in place.
3 Use a hacksaw to cut the pipe.
4 Loosen the vice and remove the piece of pipe.

Changing a fuse in a plug
1 Check that the power is off.
2 Take out the old fuse and dispose of it carefully.
3 Check that the new fuse is the correct voltage.
4 Push it into the fuse holder.
5 Turn on the power and check that the plug works.

Climbing up a ladder
1 Make sure you are wearing appropriate shoes or boots.
2 Position the ladder securely. (Ask someone else to hold it, if possible.)
3 Test each step as you climb.
4 Make sure your feet are in the middle of the steps.
5 Move the ladder every time you have to reach something far away.

D Write sentences giving instructions

- Remind students that they should use the imperative and write their instructions from memory.

- After they have written their instructions, they should compare them with the ones on pages 218 and 228.

CLOSURE

- Review the verbs for giving instructions.

Lesson 4: Describing and explaining things that are happening now

Objective

- to focus on language for describing actions in progress

Language

- present continuous for actions in progress
- present continuous for giving reasons

Vocabulary

- words related to the energy industries

LEAD-IN

- Elicit from students what hours they work and whether they work shifts. Ask them if they ever work late. If they do, ask why.

A ◉ (CD1 T30) Listen to a conversation for detailed understanding

- Ask students to read through the statements, then listen to decide if they are true or false.

- Play the recording for students to listen and complete the task.

- Ask for answers and play the recording again, pausing at the relevant places to check.

- Ask more questions, e.g., *What equipment is faulty? What is the problem?*

Answers

1 F
2 T
3 F
4 F
5 T
6 F

Tapescript

Presenter:

Lesson 4 Describing and explaining things that are happening now

A Listen to the conversation between Bob and Ahmed. Decide whether the statements below are true or false.

Ahmed: What are you doing, Bob? You don't usually work so late.

Bob: I'm staying late tonight because there's a problem with this forklift truck.

Ahmed: What's the problem?

Bob: The hydraulic mechanism is faulty. This valve isn't working properly.

Ahmed: What are you doing now?

Bob: I'm trying to have a look at it, but it's a bit hot.

Ahmed: Do you need a hand?

Bob: It's okay, there's an engineer coming in the morning.

Ahmed: Right, Bob. See you tomorrow.

Bob: Okay, see you.

B Form the present continuous

- Ask students if they can remember Ahmed's question at the beginning of the dialogue in exercise A, and Bob's reply: *What are you doing, Bob? I'm staying late tonight because there's a problem.* Write these on the board. Elicit some other examples using the present continuous, e.g., *I'm teaching at the moment. What are you doing, Mahmoud?* (*I'm listening.*) *What's he doing?* (*He's listening.*)

- Refer students to the table and ask them to complete it in pairs.
- Conduct feedback. Check students are clear about the use of the auxiliary forms *am/is/are*. Focus on how the negative is formed.

Answers

I'm (I am) *staying* late tonight.
The valve *isn't* working properly.
Bob and Mustafa *are* repairing the system.
What *are* you *doing* now?

C Practise the present continuous

- Elicit that we use the present continuous to describe something that is happening now and ask for examples from the class about what they and their classmates are doing at the moment. Perform some actions yourself to elicit *you're writing*, *drawing*, etc. Check the time and ask what friends and family are doing at the moment.
- If necessary, drill some negative and question forms.
- Refer students to the cartoon and set the task for individual work.
- Feedback answers from individual students. Make incorrect statements about the cartoon for students to correct, e.g., *John is shouting. No, John isn't shouting. The manager is shouting.*, etc.

Answers

The manager is pointing at John's ear protectors.
The manager is shouting at John.
John is holding a hose.
The manager is wearing ear protectors.
John isn't wearing his ear protectors.

D Mime and guess actions

- Mime some actions for the students to guess, e.g., writing with a pen, filling up your car with petrol, etc. Ask *What am I doing?*
- Set the task for pairwork.
- Conduct feedback. Ask some students to mime their actions to the class.

E Ask and give reasons for an activity

- Elicit some questions and answers in the present continuous using *Why? Why are you opening the window? To make the room cooler. Why are you going out? To buy some food.* Put one pair on the board and point out the use of *do* in the answer to show purpose.
- Read the short exchanges with one of the students before setting the task for individual work.

Answers

1 **A:** Why *are you disconnecting the power?*
   B: To change the fuse.
2 **A:** Why *are you pressing that button?*
   B: *To stop the machine.*
3 **A:** Why *are you increasing the volume?*
   B: *To hear it more clearly.*

F Practise dialogues

- Ask students to practise their dialogue exchanges in pairs. Ask them to invent some more exchanges using different verbs.
- Choose a couple of pairs to read their exchanges to the class.

CLOSURE

- Revise forms of the present continuous by writing a sentence on the board and eliciting the negative and as many questions as possible, e.g., *He's talking to someone on the phone. We're going on holiday.*

Lesson 5: Giving warnings

Objective

- to practise using language for giving warnings

Language

- the imperative for warnings
- present continuous

Vocabulary

- words related to safety and safety equipment

LEAD-IN

- Review the previous lesson by asking some of the students what they or their classmates are doing, e.g., *He's taking his book out of his bag*, *I'm getting a pen.* Elicit negative and question forms.

A Differentiate between instructions and warnings

- Elicit a warning from the students by pretending to put your finger in a plug socket, e.g., *Don't put your finger in the socket!* Ask if this is an instruction or a warning. Elicit the difference, i.e., warnings are given when something bad might happen.
- Remind students that the form is the imperative.
- Set the task for individual work and pairwork checking.
- Conduct feedback. Ask for more examples of warnings and instructions.

Answers

1 W 2 I 3 W 4 W 5 W 6 I 7 W 8 I

B 🔊 CD1 T31 Listen to identify warnings in a dialogue

- Set the task for individual work.

- Play the recording for students to listen and identify the sentences.
- Conduct feedback. Ask for reasons for the different warnings given in the dialogue.
- Draw attention to the intonation range of the tone, i.e., it normally moves from very high to low.

Answers

You can't smoke in here! (because it's dangerous)
Don't go in there! (because it's for restricted personnel only)
Be careful! (there are lots of safety rules)

Tapescript
Presenter:
Lesson 5 Giving warnings
B Now listen and tick the sentences you hear.

Bob: Hey, what are you doing? You can't smoke in here! It's dangerous.
New employee: Okay, I'll go outside.
Bob: You're going the wrong way. Don't go in there! It's restricted personnel only.
New employee: Sorry, I'm new here.
Bob: Well, be careful! Always read the safety signs.

C Complete sentences to form the present continuous and warnings

- Talk briefly about hazards on an oil rig and in a workshop. Check the meanings of *supervisor*, *restricted* and *flammable*.

- Read the first situation and warning with the students. Set the completion task for individual work and pairwork checking.

- Conduct feedback. There may be several possible answers for the warnings.

Example answers

2 Somebody *is talking* when the supervisor is explaining a job. *Be quiet!*
3 Somebody *is entering* a restricted area. *Don't go in there!*
4 Somebody *is carrying* a cup of coffee and *isn't looking* where they are going. *Be careful! Don't drink in here! Watch where you're going!*
5 Somebody *is smoking* near some flammable containers. *Put out that cigarette! Don't smoke in here!*

D Practise writing and giving warnings

- Set the task for pairwork.

- Feed back answers from individuals. Check that they use the correct intonation.

- Ask students what dangerous situations they can have in their workplaces and what warnings they have to give. Ask if they have any signs or notices with warnings on.

Example answers

1 You can't come in here without PPE!
2 Put on your safety harness!
3 Slow down! Don't drive so fast!
4 Be careful! Wear your safety goggles! Put on your gloves!
5 Don't stand so close! Move away!

E Use the present continuous to describe pictures and give warnings

- Set the task for pairwork. Divide pairs into Student 1 and Student 2. Make sure students are looking at the relevant pages. Clarify that students should not look at each other's pictures.

- Conduct feedback. Ask what the situations were and elicit warnings.

Example answers

Student 2:
Don't carry heavy things like that!
Bend your legs when you lift things!
Don't climb a ladder with that pipe!

Student 1:
Don't run in here!
Don't stand too near!
Keep the paper away from the welding!

F Find and correct mistakes in sentences using language from the unit

- Explain that the exercise uses grammar from the unit. Set the task for individual work and pairwork checking. Tell students to look back at the language patterns in Lessons 4 and 5.

- Go over the grammar rules of any sentences.

Answers

1 *Don't smoke/You can't smoke/No smoking* in the workshop!
2 Why *aren't you* wearing safety boots?
3 I need a spanner *to* tighten this nut.
4 Don't forget to *turn* the computer off.
5 Pick that box up *carefully*.
6 This mobile phone *isn't* working properly.

CLOSURE

- Ask how many warnings students can remember from the lesson.

Lesson 6: Comparing temporary and permanent situations

Objectives

- to compare the present simple and continuous tenses
- to talk about temporary and permanent situations

Language

- present simple
- present continuous

LEAD-IN

- Remind students of the first lesson in the book when they studied asking for and giving personal information. Ask if they can remember where you come from and any of your details. Put on the board *I come from … . I work at … . I enjoy … .* Ask some students the questions about themselves.

- In pairs, they can ask and answer the questions and feed back to the group. Highlight the question and negative forms.

A Identify examples of present simple and present continuous tenses

- Refer students to the picture and elicit some information, e.g., *Where are they? What are they doing?*

- Elicit which tenses were used in the Lead-in (the present simple for facts and permanent situations) and the activity above (the present continuous for situations and activities in progress now).

- Set the task for individual work and pairwork checking.

- Conduct feedback.

Answers

Present simple: He comes; He works, enjoys; is
Present continuous: he's living; he is having; practising

B Study the rules for the present simple

- Read through the *Present simple* information box. If necessary, clarify the difference between a *habit* (I do it every day) and a *fact* (it is always true).

- Ask students to write in examples from the text in exercise A.

- Practise further by eliciting some personalized sentences about the students.

- Repeat the procedure with the *Present continuous* information box. If necessary, clarify the difference between a temporary situation (happening for a few hours, days or months) and an action in progress (happening now).

Answers

Present simple:
For facts: He is a technical trainer; He comes from Scotland; Today is a holiday
For habits: He works
With certain verbs: (he) enjoys

Present continuous:
For actions in progress: (he) is having coffee; (he is) practising
For temporary situations: he's living

- Elicit whether any students are doing different work from normal – to illustrate the idea of the present continuous for temporary situations, e.g., *Ahmed usually works on the rigs*, *but at the moment he's working on site at … .* If students aren't doing anything appropriate, invent a few temporary situations for your family members. Ask concept questions to establish that the situations are only temporary.

C Choose the present simple or continuous to complete exchanges

- Go through the situations with the students, eliciting whether they refer to now or something that is always true.
- Set the task for pairwork.
- Conduct feedback. Ask for reasons.

Answers

1   The room gets too hot. (always) Install air conditioning. (permanent solution)
2   The room is getting too hot. (today) Open the window. (temporary solution)
3   The valve doesn't work. (never) No, we never use it. (permanent situation)
4   The valve isn't working. (now) Check it. (temporary problem)
5   Does he work on the rig? (permanent job) Yes, but he's on holiday at the moment. (habitual part of his yearly contract)
6   Is he working on the rig. (temporary job) Yes, he'll be back next week. (his work isn't finished yet)
7   Are you using this torch? (now) Not now, but I'll need it later. (plans to use it)
8   Do you use this torch? (usually in your work) No, I've got a better one. (never)

D Put verbs in the correct form to complete a paragraph

- Set the task for individual work and pairwork checking. Do not confirm answers at this point.

E 🔊 CD1 T32 Listen to check answers

- Play the recording for students to listen and check their answers.
- If necessary, go through the reasons for each choice to clarify now, always true, and verbs that do not usually take the continuous form.

Answers

1 is studying 2 has 3 is doing 4 likes
5 shows 6 are studying 7 knows
8 is finding 9 explains 10 do

Tapescript
Presenter:
Lesson 6 Comparing temporary and permanent situations
E Listen and check your answers.

Voice: Hassan is studying to be a production engineer. He has a degree, and at the moment he is doing a training course. The course involves studying in the workshop and working on site. He likes the training that happens on site because it shows how things work in action and not just in theory. This week they are studying pigging. He knows some things already, but is finding the course useful, especially when their trainer explains about the different types of pig, and what they do.

CLOSURE

- Ask students to write some sentences that are true about themselves or a partner to show the use of the present simple and continuous, e.g., what they usually do and what they are doing now, as in the model in exercise A.
- Conduct feedback. Get some examples from the class.

Lesson 7: Talking about problems in the workshop

Objectives

- to practise using the present continuous and the imperative to describe workshop problems
- to write rules for the workshop

Language

- present continuous
- the imperative

Vocabulary

- words related to problems in the workshop

LEAD-IN

- Discuss the clothes that students are wearing now and the clothes they usually wear at work. Do they wear uniforms, overalls and/or PPE?

A Use the present continuous to describe pictures

- Refer students to the three pictures. Set the writing task for individual work and pairwork checking.
- Conduct feedback. Ask for answers from different pairs.

Answers

Right:

2   He's wearing safety goggles.
3   He's wearing safety gloves.

Wrong:

1   He isn't wearing correct PPE.
2   He's using a broken ladder.
3   He's stretching too far.
4   He isn't wearing overalls.
5   He's smoking.

B Write rules for a workshop

- Put students in small groups and ask them to think of and write down some important rules for working in a workshop. They can also write down some rules for working in their different workplaces, e.g., an office.
- Scaffold the task if necessary by writing useful language on the board, e.g., *always*, *never*, *don't*.
- Feed back suggestions to the class. Ask if students have these rules in their workshops.

C ◆ (CD1 T33) Listen to a conversation to identify problems mentioned

- Explain that Vasily and Alan mention two problems in their conversation.
- Play the recording for students to listen for two problems.
- Conduct feedback.

Answers

1   The machine won't stop because the switch isn't working.
2   The emergency shut down isn't working either.

Tapescript
Presenter:
Lesson 7 Talking about problems in the workshop
C **Listen. What are the problems?**

Vasily: What's the problem?
Alan: It's this machine. It's overheating.
Vasily: You'd better turn it off.
Alan: I can't. The switch isn't working, so the machine isn't stopping.
Vasily: Try the emergency shut down.
Alan: That isn't working either.
Vasily: Now that is a problem! Turn off the power and restart it in ten minutes.

D 🔊 (CD1 T34) **Listen to complete the dialogue**

- Set the task for individual work and pairwork checking.
- Play the recording again, pausing after the inserted words for students to listen and check.
- Use prompts to elicit the conversation.

Answers

1 is
2 doesn't work
3 won't stop
4 Try
5 doesn't work
6 Turn off
7 restart

Tapescript
Presenter:
D **Listen again and complete the dialogue. Put the verbs into their correct form.**

Vasily: What is the problem?
Alan: The switch doesn't work, so the machine won't stop.
Vasily: Try the emergency shut down.
Alan: That doesn't work either.
Vasily: Turn off the power and restart in ten minutes.

E **Write solutions for different problems**

- Tell students you have a problem. Your TV remote control isn't working. Elicit possible solutions: *change the batteries*, *buy a new one.*
- Set the task for pairwork. Students should note down at least two solutions for each problem.
- Conduct feedback.

F **Create short dialogues about problems and solutions**

- Remind students to use the same exponents as in the dialogue, e.g., *What's the problem?*, *isn't working*, *Try … .*
- Set the task for pairwork. Monitor and advise where necessary. Ask some pairs to repeat their dialogues for the class.

CLOSURE

- Ask students to tell the class about any common problems they have, similar to those from this lesson, at their place of work. Elicit solutions from the rest of the class.

Lesson 8: Talking about the weather

Objectives

- to listen to a weather forecast
- to describe different weather conditions

Language

- present continuous

Vocabulary

- verbs and adjectives related to the weather
- size adjectives

LEAD-IN

- Ask some students if they have travelled abroad. What was the weather like? Ask what types of weather the students prefer, and why.

A Identify weather conditions

- Elicit other weather words students already know and write them on the board. Focus on the words *sun* and *rain* and elicit how we can describe the weather using these words. Elicit the use of the present continuous plus adjective, e.g., *It's raining. It's rainy.* Refer students to *The weather* information box.

- Ask students to look at the visuals and the words in the word box. Set the task for individual work.

- Conduct feedback. Ask for two descriptions for each picture, where possible. Get students to repeat the descriptions, paying particular attention to the endings *-ing* and *-y*.

- Ask students to name a part of the world where the weather conditions might be like these at the moment, e.g., *I think it's raining in England at the moment!*

Answers

1 It's raining./It's rainy. 2 It's cloudy.
3 The sun is shining./It's sunny.

4 It's snowing./It's snowy.
5 It's windy. 6 It's foggy.

B Study more weather conditions

- Read through the two columns and ask students to repeat the weather words after you. Check that students say the different vowel sounds correctly.

- Set the matching task for individual work and pairwork checking.

- Conduct feedback. Elicit that *ice* and *mist* can add *-y* to become adjectives. Ask when the students last experienced these conditions.

Answers

1 ice – frozen water on the ground 2 mist – thin fog 3 lightning – electricity discharges from a cloud 4 thunder – the sound of electricity discharged from a cloud 5 sleet – rain and snow together 6 hail – ice falling from clouds

C (CD1 T35) Listen to a weather forecast to answer questions

- Ask the weather forecast for today.

- Explain they are going to listen to a weather forecast for the UK in winter. Ask them to predict what might be included in the forecast. Write their predictions on the board.

- Set the task for individual work and pairwork checking.
- Play the recording for students to listen and answer the questions. Then play the recording again for them to check.
- Conduct feedback. Write the words *weather conditions*, *fog*, *frost* and *winds* on the board and ask students which adjectives were used with them: *poor*, *thick*, *heavy* and *strong*. Also ask which word meaning *dangerous* was used in the forecast.

Answers

1 There was cloud and thick fog. The temperatures were near freezing.
2 It's not so cold and there should be some sunny intervals.
3 It's snowing in Aberdeen.
4 Frost tonight could cause icy roads, which might be dangerous.

Tapescript
Presenter:
Lesson 8 Talking about the weather
C **Listen to the weather forecast and answer the questions.**

Voice: Well, we've had some very poor weather conditions in Britain this week due to the cold front that's come in from the Atlantic. There was cloud and thick fog in many parts of Scotland yesterday, and temperatures near freezing. Most areas are quite cold today, but there should be some sunny intervals later, although there could be rain and sleet showers – particularly in the North. In fact, at the moment it's snowing in Aberdeen. Rain over the hills will clear later, but we're expecting a heavy frost, and watch out for icy roads – they could be hazardous tomorrow morning. Temperatures down to minus 3 tonight. It's looking better at the weekend, but strong winds will make it feel cold. That's all from me this morning. Drive carefully and have a good weekend.

D Complete the sentences

- Set the task for individual work and pairwork checking.
- Conduct feedback. Ask students to repeat the sentences for pronunciation practice.

Answers

1 cloud; thick fog 2 snowing 3 cold
4 Rain 5 Strong winds 6 icy roads

E Interpret weather icons

- Refer students to the weather icons and set the discussion task for pairwork. Encourage them to guess the icons they don't know.
- Ask for suggestions from different pairs.

Answers

1 It is snowing. 2 It is sunny. 3 There is fog or mist. 4 It is raining. 5 There is a thunderstorm. 6 There is ice or hail.
7 There is an easterly wind. The wind speed is 15 knots. 8 The temperature is 35°C.
9 The temperature is −1°C.

F Discuss the effects of weather on oil industry operations

- Check the meaning of *helicopter flights* and *crane operations*. Set the discussion task for pairwork.
- If students have no experience of the operations in the list, ask them to think of other situations and events that may be affected by the weather, e.g., public transport, sports events, etc.
- Ask students to discuss their ideas in an open group. Are there any other operations that require certain weather conditions?

CLOSURE

- In pairs, students test each other on the weather words from the lesson by drawing symbols or giving definitions to elicit the correct descriptions.

Lesson 9: Talking about crane controls

Objectives

- to learn vocabulary for the controls of a crane
- to describe the contents of a crane

Vocabulary

- words related to crane controls

LEAD-IN

- Elicit from students instructions on how to start and stop a car.

- Elicit the word *crane* and ask what a crane is used for. If they are unsure, let them check in the glossary.

- Ask how many students can drive a crane, how many can drive a car and how many can drive both. Elicit whether driving a crane is similar to driving a car. Elicit or point out that we use the words *drive* and *operate* for a crane. Some cranes are mobile and have wheels, whereas others are fixed.

A Discuss crane and car parts

- Ask students to name as many parts used in a car as they can. If necessary, clarify the words *accelerator*, *wheels*, *steering wheel* and *brake*.

- Students name as many crane parts as they can. Elicit what these parts do. Focus on the difference between a *boom* and a *jib*: a boom is the steel arm that projects upwards. Some cranes have a jib at the end of the boom. This is a structure that the load is suspended from.

- Refer students to the word box.

- Set the task for individual work to check their understanding of the key vocabulary.

- Conduct feedback. Discuss answers as a group.

Example answers

1. steering wheel
   wheels
   accelerator
   brake
2. steering wheel
   wheels
   accelerator
   boom
   brake
   jib
   emergency stop button
   lever
3. steering wheel
   wheels
   accelerator
   brake

B Understand how the controls of a crane work

- Refer students to the picture of the crane and elicit the different parts.

- Remind students of the meanings of *vertical* and *horizontal* and check meanings of *speed* and *load*.

- Set the comprehension task for individual work and pairwork checking.

- Read the text aloud while students read it again and ask for answers to the questions.

Answers

1 No. A different set of controls are responsible for the movement.
2 The accelerator makes the crane go faster. (*increases the speed of*)
3 The steering wheel controls the direction of the crane.
4 No, it controls vertical movement.
5 Yes, lever C can raise the load.
6 Press button A (the emergency stop button) if there is an incident.

C Use new words to label a diagram of a crane interior

- Refer students to the visual and the word box. Elicit what is happening.
- Set the task for individual work and pairwork checking.
- Conduct feedback. Ask students to say again what each control does.
- Ask what else they can see in the visual that has not been mentioned or labelled.

Answers

1 boom
2 jib
3 load
4 emergency stop button
5 steering wheel
6 accelerator
7 lever
8 brake

CLOSURE

- Ask students what the potential dangers of operating a crane are.

Lesson 10: Talking about instructions for crane operations

Objective

- to read, follow and give instructions for crane operations

Vocabulary

- words related to crane operations

LEAD-IN

- Review the previous lesson by asking students to name the parts and controls of a crane.

A Group similar types of verbs

- Refer students to the groups of verbs.
- Set the task for individual work and pairwork checking.
- Conduct feedback. Ask for the word with the different meaning, then get students to tell you why they chose this verb.

Answers

1 load
2 signal
3 retract
4 inch

B Match verbs with diagrams

- Set the matching task for individual work.
- Conduct feedback.
- Ask students to discuss the difference between the verbs in each group.

Answers

1 raise: lift slowly, often large things, e.g., *raise a flag*
hoist: lift with mechanical help
lift: a more general term

2 slew: move sideways (in an uncontrolled manner)
swing: move sideways through the air
slide: move when there is continuous contact with a surface

3 extend: stretch to a greater or the fullest length, e.g., *extend a radio aerial*
telescope: extend or retract as a telescope does
lengthen: more general term; to make something longer

4 stop: general term, less formal than *halt* or *cease*
cease: discontinue or stop permanently (formal), e.g., *cease operations*
halt: stop or pause, e.g., *Let's halt for lunch.*

C Use words and phrases to describe crane operations

- Ask students to talk in pairs about what can go wrong during crane operations.
- Conduct feedback. Discuss their ideas with the class. How can these things be prevented?
- Refer students to the word box and texts.
- Set the task for individual work and pairwork checking.
- Conduct feedback. Go through the collocations: *attempt a lift, lose sight of, field of vision, poor signals, take the time to* and *cause an accident.*

Answers

1 crane operators
2 load
3 swing
4 at all times
5 clear and correct
6 vision
7 careless
8 accident
9 stop
10 signals

D Answer comprehension questions about crane operations

- Set the task for pairwork.
- Conduct feedback.
- Books covered. Ask students what sort of damage crane operators can cause. Ask them what a good spotter must do. Ask for some safety tips. Students can uncover the text to check answers.

Answers

1 To lift and lower the load.
2 To help the operator to do the job safely.
3 It is necessary to see, or be in contact with, the spotter at all times.
4 The operator may mistake one signal for another and there may be an accident.

E Understand and use hand signals

- Direct attention to the signals in exercise E. Clarify any unfamiliar terms such as *inch the load*. Ask students to imagine you are a signal spotter. Make the signals in the exercise and elicit the meanings of the signals from the students.
- Refer students to the exercise and read through with them to check and reinforce the meanings. They should repeat the signals after your model.
- Set the task for pairwork. The signalling student can look at the book and record which signals were given. His/Her partner can't. They should then swap roles.

- Books open. Students check what they have written.

CLOSURE

- Discuss any other signals that students use in their jobs.

Review: Giving instructions and warnings

Answers

A

O	Z	S	J	I	B	E	R	S	U	D	L
U	E	X	T	H	N	G	U	P	S	H	I
G	X	A	I	O	S	B	T	O	H	A	G
T	P	L	L	I	T	N	W	T	J	T	H
A	C	C	E	L	E	R	A	T	O	R	T
B	O	S	V	I	E	W	B	E	N	A	N
O	S	H	E	N	R	N	T	R	M	I	I
M	I	A	R	X	I	D	C	R	A	N	N
I	O	L	O	E	N	I	N	D	B	I	G
S	N	L	N	D	G	O	N	I	T	N	R
T	B	O	I	B	W	E	F	O	G	G	Y
O	D	W	T	B	H	Z	C	I	M	E	A
T	H	U	N	D	E	R	A	C	T	E	R
A	F	I	R	D	E	H	L	L	Y	T	N
N	S	H	E	X	L	G	O	N	A	S	D

B

1 steering wheel
2 accelerator
3 lever; jib
4 spotter
5 thunder; lightning
6 mist; foggy

C

Student 1's picture:
The spotter is signalling to hoist the load.

Student 2's picture:
The spotter is signalling to lower the jib.

Assess your skills: Giving instructions and warnings

- Refer students to the self-assessment grids.

Word list

accelerator	mark
adjust	mist
avoid	operator
boom	PPE
brake	press
button	pull
careful	push
careless	raise
cease	release
cloud	remove
cloudy	retract
crane	safety
dial	scaffolding
emergency	set
extend	signal
fog	sleet
foggy	slew
fold	spotter
hail	steering wheel
hoist	strain
horizontal	sunny
icy	swing
jib	tear
leak	thunder
lever	vertical
lightning	wheel
load	windy

DESCRIBING SYSTEMS

Lesson 1: Describing systems and devices

Objectives

- to review terminology for devices and systems
- to follow a description and answer questions about pressure measurement systems

Vocabulary

- words related to electrical systems

LEAD-IN

- Elicit different types of systems that students are familiar with, e.g., *heating systems*, *alarm systems*, *electrical systems* and *pump systems*. Discuss pressure measurement systems, eliciting or explaining the importance of monitoring pressure in the energy industries. See whether students know of, or have experience of, accidents caused by pressure changes, e.g., blowouts in oil wells.

A Name devices and systems

- Set the task for individual work.
- Conduct feedback. Ask for suggestions for where these devices might be found.
- Remind students to learn the following collocations: *change a fuse*, *read a meter*, *display information*, *sound an alarm* and *regulate the flow*.
- Ask students if they often use these words in their day-to-day work.

Answers

2   generator
3   circuit
4   fuse
5   meter
6   visual display unit
7   alarm
8   valve

B Identify devices

- Set the task for individual work.
- Conduct feedback.
- If necessary, clarify the term *component* by explaining that the devices in these diagrams are all components of systems.

Answers

1   alarm
2   fuse
3   meter
4   VDU

C Understand a description of a system

- Lead in to the text on pressure measurement in pipelines by asking what can happen if the pressure is too high or too low.
- Refer students to the *Systems* information box. Let them read it carefully.
- Set the reading task for individual work. Elicit that the description is clear. It starts by listing the components of the system, then explains the role of each component.

D Answer comprehension questions about a description

- Set the task for individual work and pairwork checking.
- Elicit answers around the class. If there is any unfamiliar vocabulary, encourage students to guess its meaning (and check *alarm*, *generator*, *relay* and *operator* in the glossary).

Answers

1 There are seven. There is a monitoring device, a generator, a back-up generator, a relay, an alarm, a control panel and a visual display unit.
2 It measures the pressure in the pipeline and takes the information via the relay to the visual display unit.
3 An alarm sounds to alert the operator.
4 It initiates an emergency shutdown.
5 If there is a failure in the primary generator or when the primary generator is undergoing maintenance.
6 The purpose of the system is to measure pressure.

CLOSURE

- Books closed. Ask students to describe the pressure measurement system to you in their own words.

Lesson 2: Describing heating systems

Objectives

- to understand and give descriptions of heating systems
- to discuss the advantages and disadvantages of different systems

Vocabulary

- words related to heating systems

LEAD-IN

- Elicit what students know about heating systems. Ask for the names of the two types of system.

A Describe an open-loop heating system

- Explain that students will be reading about an open-loop system and ask what this system consists of.
- Books open. Refer students to the diagram to check.
- Discuss the vocabulary used in the diagram labels as necessary.
- Set the completion task for individual work and pairwork checking.
- Conduct feedback. Ask individual students to read different sentences.
- Ask some comprehension questions, e.g., *Are open-loop systems manual or automatic? Where are the burners? What does the indicator show? How can the operator adjust the temperature?*

Answers

1. indicator
2. hand gas control valve
3. Cold water
4. burners
5. Hot water
6. out
7. temperature
8. gas
9. burners

B Describe a closed-loop heating system

- Elicit what students know about a closed-loop system.
- Refer them to the diagram and encourage students to discuss how the system works, in pairs. Monitor and give help (information about the system), if necessary, e.g., clarify that the controller replaces the human operator.
- Direct students to the text. Set the task for individual work.
- Conduct feedback, but do not confirm answers at this point.

C 🔊 (CD1 T36) Listen to check answers

- Play the recording for students to listen and check their answers.
- Discuss differences between the students' descriptions and the one on the recording. They do not have to have exactly the same wording, but should convey the same information.

Answers

1. a controller, a sensing element and a control valve
2. the flow and sends information to the controller
3. a signal to the control valve

Tapescript
Presenter:
Unit 5 Describing systems
Lesson 2 Describing heating systems
C **Listen and check your answers.**

Voice: Open-loop systems
Open-loop systems are manual control systems. The system is controlled by an operator. The system includes a water heater tank, an indicator, burners and a hand gas control valve. Cold water enters the tank through the 'in' pipe at the bottom. The water is heated by burners below the tank. Hot water rises to the top of the tank, and leaves it by the 'out' pipe. An indicator shows the temperature of the water in the 'out' pipe. The temperature can be adjusted by the operator using the hand gas control valve to change the gas supply to the burners.

Closed-loop systems
Closed-loop systems are automatic control systems; a controller takes the place of the operator. The system consists of a controller, a sensing element and a control valve. The sensing element monitors the flow and sends information to the controller. The flow can be adjusted by the controller sending a signal to the control valve, which regulates the flow.

D **Discuss the advantages and disadvantages of two systems**

- Set the task for individual work, then pairwork. Divide pairs into Student 1 and Student 2. Make sure students are looking at the relevant pages. They should first note down their ideas to complete their table, then discuss their ideas with their partners.

- Students take it in turns to read out the bullet points provided to their partners. Emphasize that students should not look at their partners' table.

	Advantages
Open-loop systems	• Cheap. • Easy to install. • Easy to operate. • Easy to maintain.
Closed-loop systems	• Can be used in dangerous areas. • Quick and efficient. • Does not require a human operator to be present.
	Disadvantages
Open-loop systems	• Relies on a human operator at all times. • Cannot be used in dangerous areas. • Processes can be affected by slow operator reactions.
Closed-loop systems	• More complex to repair. • Danger of accidents if the controller does not register a problem. • If the instrument fails, flow operates outside preset levels.

CLOSURE

- Students discuss the advantages and disadvantages of both systems with the whole class and suggest any additional points that they came up with.

Lesson 3: Describing alarm systems

Objectives

- to listen to a description of alarm systems and practise taking notes
- to discuss emergency procedures
- to write sentences using the zero conditional

Language

- zero conditional

Vocabulary

- words related to alarms

LEAD-IN

- Elicit what would happen if there was a fire in this building. Elicit that an alarm *warns* us of something.

A Describe different types of alarms

- Ask what other types of alarms there are and where we can find them, e.g., in our homes, in our cars, at work.
- Refer students to the visuals and ask for the difference between the two alarms.

Answers

In the car, the alarm is visual. In the building, the fire alarm makes a noise.

B ◀) (CD1 T37) Listen to a description of alarm systems to answer *true/false* questions

- Set the task for individual work and pairwork checking. Tell students not to worry about understanding every word, as they will hear the text twice.
- Play the recording for students to listen and complete the task.
- Conduct feedback.

Answers

1 F
2 T
3 T

Tapescript
Presenter:
Lesson 3 Describing alarm systems
B Listen to someone describing alarm systems. Decide whether the statements below are true or false.

Voice: There are two types of alarm – audible alarms and visual alarms, although often both are used together. An audible alarm warns the operator there is a problem, and a visual one can give a more specific idea of where the problem is. Basically, an alarm is an on-off control circuit; one that uses a limit-sensing device connected to a warning device. The alarm will go off when the equipment or process is operating outside the pre-set, normal operating range, for example, if there's too much pressure or heat, or too great a quantity of something. The limit-sensing device can be a pressure, temperature, float-operated or flow-actuated switch. When the limit is reached, the switch contacts close, completing an electrical circuit that activates the alarm.

C 🔊 (CD1 T38) **Listen to complete notes**

- Read through the notes with students and elicit any answers they can remember.
- Set the listening/checking task for individual work.
- Conduct feedback. Ask individual students to read their sentences in turn. Write key words on the board to check spelling.
- Check comprehension by asking questions about the system, e.g., *What situations are outside the preset, normal operating range?* (*too much/ little pressure, wrong temperature, etc.*)

Answers

audible alarm
visual alarm
control circuit
connected to a warning device
if there is too much pressure, heat, etc.
pressure, temperature

Tapescript
Presenter:
C **Listen again and complete the notes.**
[REPEAT OF EXERCISE B]

D **Match consequences and actions**

- Tell students about one particular warning light in your car, e.g., *if the oil is low, the oil warning light comes on.* Elicit similar examples with the same structure from students, e.g., *if the car is short of petrol ... .*
- Write one example on the board and indicate that the first part (the *if*-clause) is the action and the second part is the consequence. Elicit whether these consequences always happen as a result of these actions. (Yes.) Ask students to name the tenses in both parts of the sentences. Explain that we call this structure the zero conditional.

- Books open. Refer students to the word box and the table.
- Set the matching task for individual work and pairwork checking.
- Conduct feedback.
- Read through the information boxes with the students to check understanding.

Answers

If there is a fire in the building, *the alarm sounds.*
If the primary system fails, *the back-up system comes online.*
If the switch is in the OFF position, *the lights are red.*
If the valve is closed, *the oil flows to the overflow tank.*

E **Complete zero conditional sentences**

- Set the task for pairwork. Point out that students should use their own ideas for sentence 4.
- Conduct feedback. Ask for examples from different pairs.

Example answers

1   If the alarm goes off, we *leave the building.*
2   If *there is a fire,* the alarm goes off.
3   If there is an accident, *we call the hospital.*
4   If *it's too hot, the warning light comes on.*

CLOSURE

- If students need extra practice, ask them to work in pairs and write the first part of three conditional sentences for their partners to finish. They should then write the second parts for their partners to start.

Lesson 4: Describing how electrical systems work

Objectives

- to read and write about different electrical systems
- to ask questions about systems and procedures using the zero conditional

Language

- zero conditional for questions

Vocabulary

- words related to electrical systems

LEAD-IN

- Review the previous lesson by asking students if they have seen or heard any alarms since the last lesson and to tell the class about them.

A Discuss diagrams showing electrical systems

- Ask students what they know about electric circuits.
- Set for pairwork. Refer them to the two diagrams and ask them to discuss how they work.
- Conduct feedback. Ask for suggestions from different pairs.

B Match a description with the correct diagram

- Ask students to read the description and decide which diagram it describes. Elicit reasons for their choice.
- Students practise reading the description aloud in pairs. Then ask them further comprehension questions, e.g., *what does the circuit consist of? How are the parts of the circuit connected? What happens when the switch is in the ON position? What happens when it is in the OFF position?*

Answer

Diagram 1

C Complete sentences with information from a description

- Check students remember what the zero conditional is.
- Set the completion task for individual work.
- Conduct feedback. Ask for answers from the whole class.

Answers

1 The lamp goes on if *the circuit is closed*.
2 The lamp goes off if *the circuit is broken*.

D Describe another basic electrical system

- Put students in pairs and ask them to work together to write a similar description for the other diagram. Remind them to use similar phrases and exponents as in the first description, e.g., *It consists of ... .*
- Conduct feedback. Ask some pairs to read their descriptions to the class.

E Complete sentences with information from a description

- Set the completion task for individual work.
- Conduct feedback. Ask for examples from the class, but do not confirm whether they are right at this point.

F 🔊 (CD1 T39) Listen to check answers

- Play the recording for students to listen and check their answers.
- Ask students to write the correct answers in their books, if they had mistakes.
- Practise using negatives in the zero conditional by giving students the first parts of a sentence to elicit the second, e.g., *I forget new English words if I don't ... . My son doesn't go to school if ... . My car doesn't run well if ... . My security alarm goes off if I ... .*

Answers

1 Lamp A goes on if switch A is in the ON position.
2 Lamp B goes on if switch A and switch B are in the ON position.
3 Lamp B doesn't go on if switch B is in the ON position, but switch A is in the OFF position.

Tapescript
Presenter:
Lesson 4 Describing how electrical systems work
F Listen and check your answers.

Voice: Lamp A goes on if switch A is in the ON position. Lamp B goes on if switch A and switch B are in the ON position. Lamp B doesn't go on if switch B is in the ON position, but switch A is in the OFF position.

G Ask questions using the zero conditional

- Ask students zero conditional questions about the two diagrams, e.g., *What happens if switch A is in the ON position?* Focus on correct word linking and pronunciation.
- Write the question on the board. Use the *What happens if* question to ask students about something simple in a car, e.g., *What happens if I don't put enough water in the car?* Elicit similar questions.
- Set the task. Ask students to find some diagrams from previous lessons and to ask and answer questions using the zero conditional, as in the example.
- Conduct feedback. Ask some pairs to repeat their questions and answers for the class. The class should guess which diagrams they are talking about.

CLOSURE

- Elicit the names for parts of an electric circuit and a simple description of the two covered in this lesson.

Lesson 5: Describing electrical systems

Objective

- to describe an electrical system using correct terminology

Vocabulary

- verb-noun collocations related to electrical systems

LEAD-IN

- Review the previous lesson briefly by asking a student to draw a diagram of one of the basic systems on the board. Elicit what the circuit consists of and how it works from the rest of the class.

- Revise the zero conditional by asking students to complete sentences associated with the diagram.

A Study words associated with electrical systems

- Books closed. Draw some items from the legend on the board and ask if students can say what they are. Ask if they know any more and get them to draw them on the board if they do for the class to name.

- Books open. Refer students to the legend and compare them with the symbols on the board.

- Read through the words with the students and check comprehension.

- Set the labelling task for pairwork.

Answers

Clockwise from top left:
fuse; switch; bell; bell; battery

B Discuss a diagram

- Ask pairs to discuss how the circuit in the diagram functions.

- Conduct feedback. Draw the diagram on the board and elicit ideas from the class.

C Match sentence halves to make correct collocations

- Write *The bell sounds the alarm* on the board and elicit that *sound the alarm* is a collocation. We cannot say *makes the alarm* or *tells the alarm*.

- Explain that in the following exercise students are going to look at some more useful collocations for describing electrical systems.

- Set the task for pairwork.

- Conduct feedback. Read the first part of the sentence and ask the class for the second. Write the collocations on the board as you go and clarify new items such as *flick*.

- Exercise covered. Say some verbs such as *charge*, *blow*, *operate*, *make*, *flick*, *ring* and *sound*. Elicit the nouns that they collocate with. Students can also test each other in this way.

Answers

1 This button sounds the buzzer.
2 How do you ring the bell?
3 You need to charge the battery.
4 The circuit is completed where it makes contact.
5 Don't flick that switch!
6 The device is operated by a push button.
7 The machine stopped working when it blew a fuse.

D Draw and describe a basic electrical system

- Set for pairwork. Explain the task. Students should first draw a basic electrical system in the top box. They should ensure that their partners cannot see what they have drawn.

- They then describe the system to their partners, who must draw what they hear in the second box. The student who is describing can watch what their partner is drawing and correct them if they make a mistake – in English!

- Students compare the original drawings and their partners' copies.

- Conduct feedback. Ask some students to describe their systems to you. You draw on the board exactly what they say.

- Elicit from the whole class how the systems you have drawn on the board work.

CLOSURE

- Review items from the legend in exercise A and the collocations in exercise C.

Lesson 6: Using adverbs of frequency

Objective

- to use adverbs of frequency to talk about valves and study habits

Language

- adverbs of frequency

Vocabulary

- types of valves

LEAD-IN

- Review the previous lesson by asking which verbs collocate with *a buzzer, a bell, a fuse, a circuit, contact, a battery* and *a switch*.

- Draw the items from the legend and elicit the meanings/names.

A Label a diagram with names of different types of valves

- Refer students to the diagram of a pump on page 64 (Unit 3, Lesson 10). Elicit that the diagram includes a type of *ball valve*. Elicit the names of other valves from the students and put them on the board. If students cannot name any, direct them to look up *valve* in the glossary for examples.

- Books open. Explain that the symbols represent different valves. Note that symbols for valves tend to vary, so students may have seen alternative symbols, e.g., for the ball valve. Check how many valves shown in exercise A are on the board.

- Read through the names of the valves and ask for examples of when these valves are used. Note that the uses of butterfly and check valves are explained in the glossary.

- Set the task for individual work and pairwork checking.

- Conduct feedback. Draw the system on the board and elicit names from the class.

Answers

Top row:
hand-operated control valve; ball valve; three-way valve; gate valve; normally-closed valve

Bottom row:
butterfly valve; three-way valve; butterfly valve; ball valve

B Rank items in order of frequency

- Elicit what can go wrong with a valve, e.g., *they can stick/leak.*

- Explain the task by reading through the situation with students.

- Set the ranking task for pairwork.

- Conduct feedback with the whole class.

Answers

1st:	the first three-way valve
2nd:	the gate valve
3rd:	the left butterfly valve (the order of 2 and 3 can also be reversed)
4th:	the second three-way valve
5th:	the right butterfly valve
6th:	the normally-closed valve

C Order frequency adverbs

- Set the task for pairwork.

- Conduct feedback. Elicit answers from the class onto a diagram on the board.

- Give students some examples of things you do or don't do regularly, using frequency adverbs. Elicit examples from them. Put some activities on the board, e.g., *use the computer*, *work late*, *wear overalls*, *travel by plane*, etc. Elicit more sentences. Point out where we normally put the frequency adverb in the sentence, i.e., before the main verb unless it is the verb *be*, when the adverb goes after the verb, e.g., *Yuri always leaves early*. But *Yuri is always late*.

Answers

never
rarely/seldom
seldom/rarely
occasionally
sometimes
often
frequently
usually
always

D Practise using frequency adverbs

- Explain that you are now going to find out some information about the students' English language learning.

- Ask students to look at the short exchanges in the speech bubbles. Act them out with one student. Ask another student the first question to elicit a true answer. Then ask the follow-up question.

- Set the interview task for pairwork. Ensure that students understand that they should ask for details and note their partners' answers down on the table.

- Conduct feedback. Ask individual students to give the class some information about their partner from their interview notes. If students need more practice, change the pairs or write more interview questions about holidays, freetime activities, etc.

E Discuss ways to improve English outside the classroom

- Write some information from the last activity on the board, e.g., *Ahmed often reads things in English. He sometimes buys English magazines and occasionally buys English books.*

- Set the discussion task for pairwork. Monitor and contribute to help the discussions along.

- Conduct feedback. Ask for different suggestions and open this up to class discussion of the suggestions made.

- Students can write their ideas, using the imperative as bullet points.

Example answers

Read more grammar books at home.
Listen to the news on the BBC World Service.
Watch more English language films.
Set up an English social club.
Read articles and blogs on the Internet. (http://freespace.virgin.net/alan.foum/ is a comprehensive web directory which lists sites connected with the energy industries.)
Keep a vocabulary notebook.

CLOSURE

- Give out a slip of paper to each student. Ask students to each write a true sentence about themselves and their work using a frequency adverb, then give in the paper. Read out the sentences and get the class to guess who wrote them.

Lesson 7: Using process and instrument drawings

Objectives

- to review vocabulary for parts of systems
- to learn abbreviations and symbols for parts of systems

Vocabulary

- devices and parts of systems

LEAD-IN

- Write the abbreviations *P & IDs* on the board and ask if students know what the letters stand for. Then write the full words, i.e., *Process and Instrument Drawings*. Elicit what these are. Ask students if they can draw any on the board and explain them.

A Match P & IDs and devices

- Refer students to the set of symbols and read through the names for students to hear the pronunciation. Allow students a few minutes to study them. Pre-teach or check unfamiliar or forgotten words in the glossary, e.g., *drain*, *condenser* and *filter*.

- Ask students to cover the symbols and draw a random set on the board. Elicit the names. Ask students to test each other in pairs. They take it in turns to name and draw.

- Ask students to look at the set of descriptions. Set the task for individual work and pairwork checking.

- Conduct feedback by asking individual students to name the item and draw the correct symbol on the board.

- Write the following verbs on the board: *provides, removes, stores, heats, cools, stops, redirects, allows, warns* and *moves*.

- Books covered. Draw a symbol on the board and elicit the name and what it

does. Students can use the verbs as prompts to give the descriptions. Again, they can test each other in pairs.

Answers

1 electric motor
2 filter
3 floating roof tank
4 heater
5 cooler/condenser
6 butterfly valve
7 four-way valve
8 drain
9 level alarm
10 compressor/blower/pump

B Study standard abbreviations

- Write the letter *T* on the board and tell students that this is an abbreviation. Ask if they know or can guess what it stands for (*tank*). Explain that abbreviations are used to help clarify diagrams. Elicit any other standard abbreviations they know and put them on the board.

- Set the task for pairwork. Divide pairs into Student 1 and Student 2. Make sure students are looking at the relevant pages. Students ask and answer questions to complete their tables. Warn students that some abbreviations are used for two different devices, e.g., *P* and *T*.

- Conduct feedback. Ask for full words and abbreviations around the class.

Answers

1

tank	T	compressor	K
pump	P	filter/strainer	S
heat exchanger	E	alarm	A
controller	C	drain	D

2

pressure	P	speed	S
furnace	F	recorder	R
level	L	indicator	I
temperature	T	motor	M

C Work out the meaning of abbreviations

- Set the task for pairwork. Students should guess and write down the meanings of the abbreviations. Students compare their answers with another pair.

- Conduct feedback. Do not confirm whether they are correct at this point.

D ◀) CD1 T40 Listen to check answers

- Play the recording for students to listen and check their answers.

- Conduct feedback. Ask where these devices can be found.

Answers

1 level indicator
2 temperature indicator
3 pressure alarm
4 level alarm
5 temperature recorder
6 flow recorder
7 pressure recorder indicator
8 speed indicator controller
9 flow indicator controller

Tapescript
Presenter:
Lesson 7 Using process and instrument drawings
D Listen and check your answers.

Voice: 1 level indicator
2 temperature indicator
3 pressure alarm
4 level alarm
5 temperature recorder
6 flow recorder
7 pressure recorder indicator
8 speed indicator controller
9 flow indicator controller

CLOSURE

- Ask students which devices covered in this lesson they are familiar with and come across in their jobs.

Lesson 8: Using tag numbers

Objectives

- to practise listening for very specific information – tag numbers and abbreviations
- to sequence and write a description of a system

Vocabulary

- words related to tanks, pressure recorders and system problems

LEAD-IN

- Write some abbreviations on the board to review the previous lesson.
- Ask students how many symbols they can remember and name.

A Identify symbols in a diagram

- Write the words *tag numbers* on the board and ask students if they know what these are and why they are used.
- Write *PR427* on the board and elicit what the letters stand for (pressure recorder, series 427). Do the same with *T-3501* and *T-4501*. Elicit that these are two different tanks. Write *3501-MV-028* and see if students can work out that this is a manual (hand-operated) valve on tank 3501.
- Refer them to the *Tag numbers* information box.
- Ask students to look at the diagrams and name any parts they can, e.g., *tanks*, *level gauge*, *alarm* and *valves*.

B (CD1 T41) Listen to label a diagram

- Explain that students are going to hear a description of the diagram using tag numbers. They probably won't be able to label everything in the diagram, so they should just try to label the tanks (labels 3 and 6) and draw in the hose that connects the tanks. Elicit or clarify the meaning of *hose*.
- Play the recording for students to listen and label the tanks.
- Write the tag numbers from the description on the board. Students practise listening to and saying them. Model the numbers. Students repeat.
- Play the recording again. Students fill in the tag numbers. Conduct feedback.

Answers

1 level gauge 4501-LG-102 **2** (manual) oil skim nozzle valve 4501-MV-006 **3** T-4501 **4** N4 **5** hose **6** T-3501 **7** manual valves 3501-MV-005 (on the left) and 3501-MV-028 (on the right)

Tapescript
Presenter:
Lesson 8 Using tag numbers
B Listen to the description of the diagram.

Voice: The T-4501 is a water storage tank located on the upper deck. Water is pumped to T-4501 from tank T-3501 located on the lower deck. To fill T-4501, a hose is connected from tank fill nozzle N4 on the bottom of T-4501 to manual valve 3501-MV-028 below. This valve must be open when you begin to fill the tank and also the one to the left of it: 3501-MV-005. You have to open the oil skim nozzle valve 4501-MV-006 at the top of the storage tank and continue to fill the tank up until water overflows from the skim nozzle. After that, you close the valves and disconnect the hose. The level of the water needs to be marked on level gauge 4501-LG-102.

C 🔊 (CD1 T42) **Listen and match sentence halves**

- Students will hear a description of a pressure-monitoring system which deals with changes in pressure in the pipeline. Pre-teach *fail*, *back up*, *malfunction* and *override*.

- Ask students to read through the sentence halves to become familiar with the tag numbers. In pairs, see if they can guess which halves match. Play the recording for students to listen and match or check.

- Play the recording again. Pause after every sentence to check answers.

Answers

1   PR427-C automatically closes valves PV576A/B.
2   The PR monitoring systems are used to monitor pressure.
3   PR427-C2 initiates if PR427C malfunctions.
4   This leads to a reduction in the flow, and lowers the pressure.
5   The PR427-C has a back-up system PR427-C2.
6   The PR427 series are suitable for measuring high pressures.
7   When the pressure reaches a preset high-high level, the alarm PR427-A alerts the operator.
8   The operator also has emergency override PR0993 if both PR427-C systems fail.

Tapescript
Presenter:
C Listen to a description of another system. Match the sentence halves.

Voice: The PR monitoring systems are used to monitor pressure. The PR427 series are suitable for measuring high pressures. When the pressure reaches a preset high-high level, the alarm PR427-A

alerts the operator. PR427-C automatically closes valves PV576A and B. This leads to a reduction in the flow, and lowers the pressure. The PR427-C has a back-up system PR427-C2. PR427-C2 initiates if PR427-C malfunctions. The operator also has emergency override PR0993 if both PR427-C systems fail.

D Rewrite a description in the correct order

- Set the task for individual work. Students write out each sentence in full.
- Conduct feedback.

Answers

2   The PR monitoring systems are used to monitor pressure.
6   The PR427 series are suitable for measuring high pressures.
7   When the pressure reaches a preset high-high level, the alarm PR427-A alerts the operator.
1   PR427-C automatically closes valves PV576A/B.
4   This leads to a reduction in the flow, and lowers the pressure.
5   The PR427-C has a back-up system PR427-C2.
3   PR427-C2 initiates if PR427-C malfunctions.
8   The operator also has emergency override if both PR427-C systems fail.

E Check answers

- Set for pairwork. Students check their descriptions with a partner.
- Then check answers with the whole group.

CLOSURE

- Put the words *fail*, *back up*, *initiate*, *malfunction* and *override* on the board with tag numbers *PR427*, *PR099*, *PR427-C*, *PV576A/B* and *PR427-C2*. Students describe the system using these prompts.

Lesson 9: Giving definitions

Objectives

- to revise vocabulary
- to practise writing definitions of people, places and equipment in the oil industry

Language

- defining relative clauses
- relative pronouns

Vocabulary

- words in definitions

LEAD-IN

- Elicit definitions of different equipment and jobs in the energy industries by asking questions, e.g., *What is a jughustler? What are geophones?*

A Use relative pronouns to give definitions

- Write on the board *A dictionary is a book. A dictionary tells us the meanings of words.* Elicit how we can combine the sentences into one. Prompt with the word *which*, if necessary, e.g., *A dictionary is a book which tells us the meanings of words.*

- Elicit that the words *who/which/where* are relative pronouns. Explain that using them in this way makes a relative clause.

- Read through the *Relative pronouns* information box with students.

- Ask students to read the rule box. Set the task for individual work.

- Conduct feedback. Ask why students chose a particular word.

Answers

1 which/that
2 who
3 where

B Combine sentences using relative clauses.

- Read through the sentences. Set the task for individual work and pairwork checking.

- Conduct feedback. Ask students to read out the complete sentences.

Answers

1 Sarah is a lead engineer who works on an oil rig.
2 A drain is a device which/that allows water to leave a tank.
3 This is the trainer who/that teaches English for the oil and gas industry.
4 A heater treater is a piece of equipment where water and oil are separated.
5 This is a cheap system which/that is easy to install.

C Complete definitions

- Write the beginnings of some definitions on the board, e.g., *Dubai is a city where ... . Ahmed is a person who ... . A Ferrari is a car that ...* . Ask the class to complete them orally.

- Refer students to the task. Set it for individual work. Stress that they should complete the sentences with their own ideas.

- Do not go through the answers at this point.

D Compare definitions

- Ask students to compare their definitions with a partner and decide which they think is the best.

- Conduct feedback. Ask pairs to read their definitions to the class.

- Ask students to work in pairs again to write the beginnings of six more definitions like those in exercise C. They should then swap their starters with another pair and complete them.

- Conduct feedback. Go through the different definitions with the whole group.

Answers

1   ... heads the drilling crew and operates the drilling machinery.
2   ... does routine cleaning and maintenance.
3   ... where oil wells are drilled from.
4   ... that controls or regulates the flow of a liquid or gas in a pipe.
5   ... cables with detectors that pick up sound waves from under the ground and the ocean.

E Write definitions for equipment and jobs

- Refer students to the visuals of different equipment and jobs. Set the task for pairwork.

- Conduct feedback.

Answers

1   A valve is something that controls or regulates the flow of a liquid or gas in a pipe.
2   A diver is a person who/that works or goes under water.
3   Geophones are instruments which/that detect sound waves under the ocean or earth.
4   A crane operator is a person who/that drives/operates a crane.
5   An electric drill is a tool which/that is used to make holes.
6   A fuse is a device that forms part of an electric circuit and protects it from excessive current.

CLOSURE

- Students take it in turns to give a definition of a person, place or thing. The class have to guess what it is.

Lesson 10: Measuring flow

Objective

• to read and compare descriptions of different flow meters

Vocabulary

• words related to measuring flow

LEAD-IN

• Review definitions by giving some items for students to define and by giving definitions to elicit the names of the items.

A 🔊 (CD1 T43) Listen to identify words and definitions

• Refer students to the word box and check meaning of the terms. (Apart from *back-up system*, all the words are in the glossary if you would like students to practise looking them up and giving definitions.)

• Play the recording for students to listen and identify which words are defined.

• Conduct feedback.

Answers

1   tank
2   overflow
3   rotor

Tapescript
Presenter:
Lesson 10 Measuring flow
A **Listen to the definitions and guess which words in the box they are describing.**

Voice: 1   It's a container that's used to hold liquid or gas. It's usually made of metal, plastic or fibreglass.

2   It's an instrument that allows excess liquid to escape.
3   It's part of a mechanical or electrical system that rotates.

B 🔊 (CD1 T44) Listen to complete sentences

• Set the task for individual work and pairwork checking.

• Play the recording again for students to listen and complete the sentences with the correct words.

• Conduct feedback. Ask different students to read a sentence each.

Answers

1   A *tank* is a *container* which is used to hold *liquid or gas*.
2   An *overflow* is an *instrument* that allows excess liquid to escape.
3   A *rotor* is part of a mechanical or electrical *system* that *rotates*.

Tapescript
Presenter:
B **Listen again and complete the sentences.**
[REPEAT OF EXERCISE A]

C Read a description for detailed information.

• Pre-teach *heater treater*, *excess* and *divert*.

- Direct students to the text and ask them to pick out the tag numbers and what they refer to (tanks, pipes and valves). Clarify that the text is describing how water is stored and diverted during treatment.
- Set the reading task for individual work and pairwork checking.
- Conduct feedback. Ask individual students to read their answers.

Answers

1   To store water.
2   Yes. A is twice as big as B.
3   When A is full; when A is undergoing maintenance.
4   It closes the overflow pipe during normal operations.
5   To the next stage of water treatment.

D Read descriptions of different meters to complete a table

- Refer students to the visuals and ask if any students know what they represent. Remind students that they looked at a flow meter (rotameter) in Unit 3. Let students read the names that head the descriptions and elicit any details they know about how these meters measure flow.

- Write this group of words on the board and ask students to work in pairs to guess which words are used in which descriptions: *conduct*, *rotor*, *transmitted*, *proportional to*, *two points*, *free-spinning*, *received*, *signal*, *turns* and *throat*.

- Indicate the table at the bottom of the page. Set the task for individual work and pairwork checking.

- Conduct feedback. Ask individual students to use the information from their tables to describe how the flow is measured and what advantages and disadvantages there are. Ask if other students agree, then check by reading the relevant parts of the descriptions.

Answers

How flow is measured:
turbine: by the movement/speed of a free spinning rotor
Venturi: by calculating difference in pressure between two points
ultrasonic: by calculating the time signals take to be sent and received

Advantages/Disadvantages:
turbine: disadvantage – affects the pressure and disrupts the flow
Venturi: advantage – has little effect on downstream pressure and flow
ultrasonic: disadvantage – only used with fluids that can conduct ultrasound and have a well-formed flow

CLOSURE

- Books closed. Ask students to describe in their own words the Lincoln 9384-A/B, a turbine meter, a Venturi meter and an ultrasonic meter.

Answers

A

1   filter – A device *that/which* removes particles from a liquid.
2   flow meter – An instrument *that/which* measures the flow of liquid in a pipe.
3   floating roof tank – A large, open container *where* oil is stored.
4   level alarm – Something *that/which* gives a warning if the level of liquid is too high or too low.
5   displacer – Something *that/which* removes something from its usual position.
6   detector – Any device *that/which* receives and responds to a signal.

B **C**

Students' own answers.

Assess your skills: Describing systems

- Refer students to the self-assessment grids.

Word list

alarm	loop
back-up	mechanism
bell	meter
burner	nozzle
buzzer	outlet
circuit	overflow
compressor	preset
concise	primary
current	rarely
deck	recorder
displacer	regulate
diverted	relay
drain	reservoir
excess	rotor
filter	seldom
float	tag
frequently	treater
fuse	turbine
generator	ultrasonic
hose	ultrasound
indicator	unidirectional
interact	variable
laminar	visual display unit

TALKING ABOUT SAFETY

Lesson 1: Talking about parts of the body and injuries

Objective

• to learn vocabulary for parts of the body and different injuries

Vocabulary

• words related to the body and injuries

LEAD-IN

• Revise body parts by discussing common injuries and seeing if students can remember the pie chart of percentages of work injuries in Unit 2, Lesson 3.

A Identify parts of the body

• Refer students to the word box and read through it.
• Check students understand *torso* and *limbs*. Ask them to group the words in pairs.
• Conduct feedback by asking different pairs to give one word each from a box.
• Set the labelling task for individual work and pairwork checking.
• Conduct feedback. Pay attention to the pronunciation of the words with silent letters, i.e., *wrist*, *knee* and *thumb*.
• Ask students to test each other by pointing to different parts of their bodies to elicit the correct words.

Answers

Torso: back; shoulder; chest
Head: neck; ear; face; nose; mouth; eye
Upper limbs: thumb; finger; elbow; hand; arm; wrist
Lower limbs: toe; ankle; leg; foot; knee

B Label a diagram

• Refer students to the diagram. Ask them to label it with the words in the word box in exercise A.

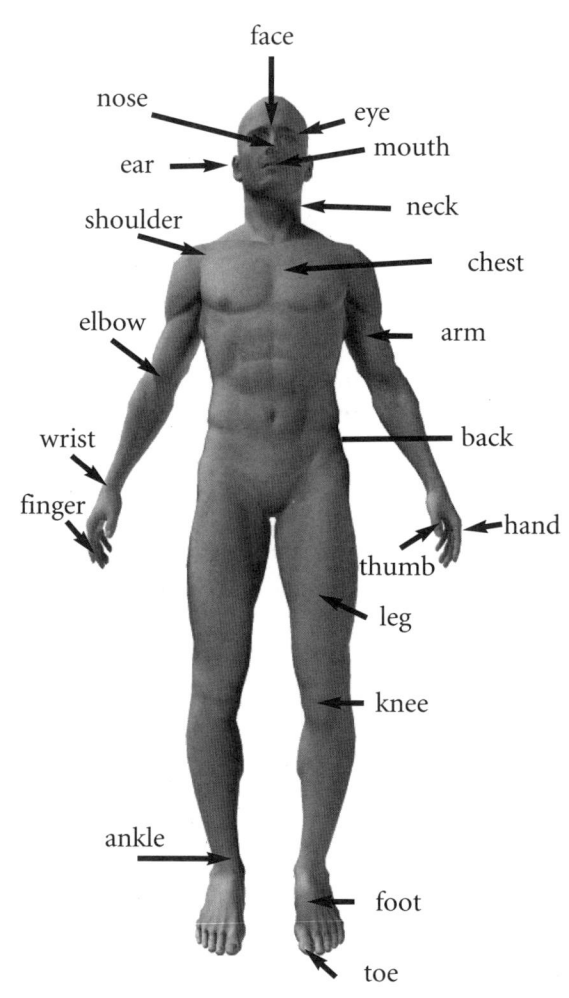

C Talk about different injuries

- Refer students to the dictionary definitions and read through them to model the pronunciation. Explain that these are the medical terms and the words that doctors use. If they are talking about an injury with colleagues and friends, they will usually say *he broke his leg* and *he cut his finger*.

- Set the task for pairwork.

- Conduct feedback. Elicit which verbs we can use when talking about an injury, e.g., *cut, break, bruise, sprain, hit, hurt* and *dislocate*. Give an example of an injury you had in the past, e.g., *I broke my ankle in a football match!* Ask for some examples from the students to show the collocations, e.g., *I hit my head; I cut my arm*, etc. For extra practice, ask each student to write down one sentence about an old injury. They should tell a partner, who must guess when and how it happened.

Answers

1 a fracture 2 a laceration 3 a contusion/ a sprain

D 🔊 (CD2 T1) Listen for specific information

- Set the task for individual work and pairwork checking.

- Play the recording for students to listen and circle the body parts Alan hurt.

- Conduct feedback.

Answers

He hurt his head, right wrist, right thumb, back and right arm.

Tapescript
Presenter:
Unit 6 Talking about safety
Lesson 1 Talking about parts of the body and injuries
D Alan was in an accident. Listen and circle the body parts he hurt.

Alan: I've had one accident at work, quite a nasty one actually. It was on New Year's day – the first of January, early in the morning. I slipped on a wet floor in the workshop and fell. I hit my head on the corner of the work bench as I fell and cut my head – it was quite a deep laceration. Then I landed badly on my hand and broke my right wrist and sprained my right thumb. I also managed to hurt my back, and I got some nasty bruises on my right arm.

E 🔊 (CD2 T2) Listen for specific information

- Set the task for individual work and pairwork checking.

- Play the recording again for students to listen and complete the accident report form. You may need to play it twice or pause the tape at points to give them time to note down the answers.

- Conduct feedback. Do not confirm whether students are right at this point.

- Play the recording and stop frequently to confirm answers.

Answers

Date and time of accident: 1st January, early morning
Place of accident: the workshop
Reason for accident: wet floor
Injuries: cut head; broken wrist; sprained thumb; hurt back; bruised arm

Tapescript
Presenter:
E Listen again and complete the accident report form.
[REPEAT OF EXERCISE D]

CLOSURE

- Ask students to tell the class about any accidents that have happened recently at their workplaces or in their families. What happened? Why? What were the injuries?

Lesson 2: Talking about PPE and safety equipment

Objective

• to review vocabulary for PPE and safety equipment

Vocabulary

• words related to PPE and safety equipment

LEAD-IN

• Review parts of the body by writing the first letters of ten parts on the board and eliciting the full words from students.

• Elicit some injuries by indicating the different parts of the body and asking what you can do to them, e.g., you can break a leg, you can sprain a wrist, you can cut your face, you can dislocate your shoulder, etc.

A Complete protection signs

• Refer students to the signs and ask where in the workplace you can see signs like these.

• Set the task for individual work.

• Conduct feedback.

Answers

2 head
3 ear
4 foot
5 eye
6 hand

B Identify PPE

• Exercise covered. Elicit different types of protective clothing. Put any answers on the board.

• Uncover the exercise and check words in the box against any suggestions on the board.

• Get students to look at the question and answer in the speech bubbles and

model them for students to repeat. Set the task for pairwork.

• Conduct feedback. Ask some pairs to repeat their exchanges to the class.

• Ask which of these items students have to wear or use, and why.

Answers

2 They are overalls. They are to protect the body.
3 They are safety goggles. They are eye protection.
4 They are safety boots. They are to protect your feet.
5 They are safety gloves. They are to protect your hands.
6 They are ear protectors. They are ear protection.
7 It is a mouth and nose/face mask. It is to protect your mouth and stop you breathing in fumes.
8 It is a hard/safety shoe. It is foot protection.
9 It is a face mask. It is to protect your eyes and face.

C Identify safety equipment

• Elicit items of safety equipment that the students know of or have in their workplaces. Ask what they are for.

• Set the matching task for individual work and pairwork checking.

• Conduct feedback. Ask *What's a machine guard? It's a … .*

Answers

1 machine guard – An attachment or a covering on a machine to protect the operator or a part of the machine.
2 first-aid kit – A box containing medicines and bandages to treat injured people.
3 fire extinguisher – A metal container containing water or chemicals for stopping fires.
4 safety barrier – A type of gate or fence that stops people going somewhere.
5 flare – A signal that produces a bright flame to attract attention.
6 life buoy – An inflatable device that floats, which you throw to someone in the water.
7 lifeline – A rope used for saving people in danger, especially at sea.
8 safety harness – A set of bands worn to hold someone, or stop them from falling.
9 safety cone – Orange and white objects that are placed around an area to stop people going there.

D Identify and discuss safety equipment

- Set the task for pairwork.
- Feed back answers and suggestions from different pairs. Discuss with the whole class the different situations when these items of equipment might be necessary. Ask if students have ever had to use them while working. Elicit where and in what situations the equipment might be used in everyday life (outside their work).

Answers

1 machine guard 2 safety harness 3 lifeline
4 fire extinguisher 5 safety barrier
6 safety cone 7 mask 8 flare 9 life jacket

E ◉ (CD2 T3) Listen to identify items that are described

- Set the task for individual work.
- Play the recording for students to listen and identify the different items of equipment.

- Conduct feedback. Ask how the people described the items. Use prompts if necessary, e.g., *colour*, *material* and *place*.

Answers

1 fire extinguisher
2 safety cones
3 mask
4 life jacket

Tapescript
Presenter:
Lesson 2 Talking about PPE and safety equipment
E Listen and identify the equipment above that people are discussing.

Voice: 1 It's a sort of container, often red, that hangs on the wall and has water, foam or chemicals in it for putting out fires.
2 They're orange and white objects made of plastic. You see them on roads or around any area that is dangerous, to stop people going there.
3 What's the name of that thing you wear on your face? It's something that's made of rubber or plastic, and stops you inhaling poisonous fumes or gases.
4 It's something that goes around you to stop you drowning if you fall in the water.

CLOSURE

- Ask students to choose five items from the lesson and describe them to their partners for them to guess. Tell them to use *It's … .*
- When they have finished, briefly and quickly run through the items yourself, giving quick definitions to elicit answers, e.g., you wear them on your feet, etc.

Lesson 3: Using different parts of speech

Objectives

- to read for specific information
- to raise awareness of and practise word-building

Language

- adjectives, nouns and verbs

Vocabulary

- words related to safety

LEAD-IN

- Review safety equipment and protective clothing from the previous lesson.

A Read a text to complete information in a table

- Elicit the word *gloves* and ask students what different types of gloves you can have to protect the hands.
- Set the reading task for individual work and pairwork checking.
- Conduct feedback. Ask which boxes students have ticked.
- Ask students to find words in the text that mean the same as 'most important' (*principal*), 'correct for the situation'

(*appropriate*), 'something that can be thrown away' (*disposable*), 'not very strong' (*mild*), 'very dangerous' (*hazardous*), 'made of cloth' (*fabric*), 'not heavy' (*lightweight*), 'things that move easily in your hands' (*slippery*), 'very thick' (*heavy duty*), 'keep out hot and cold things' (*insulate*), 'not smooth' (*rough*). Elicit examples of other nouns these words collocate with and ask students to choose four and write sentences using the words in their notebooks to record the meanings.
If necessary, point out that we say *a pair of gloves*. Elicit what other items of clothing are referred to as *a pair of* (shoes/glasses/socks/trousers/shorts, etc.).

Answers

Type of gloves	Heat/Cold	Mild irritants	Hazardous chemicals	Carelessness	Sharp edges
disposable	✗	✓	✗	✗	✗
chemical-resistant	✗	✓	✓	✗	✗
fabric	✓	✗	✗	✗	✗
leather	✓	✗	✗	✗	✓

B Match types of gloves to different situations

- Set the task for individual reading and pairwork checking.

- Conduct feedback. Ask individual students the types of gloves they chose, and why. Ask what other protective clothing or equipment they might need in these situations.

- Ask students which gloves they use themselves, and for what sorts of situations.

Answers

1   leather
2   leather
3   disposable/chemical-resistant
4   disposable
5   leather
6   chemical-resistant
7   fabric

C Identify and use different parts of speech

- Ask students if they can remember the different parts of speech. Elicit as many as they can tell you and ask for examples. Put on the board *Noun*, *Verb* and *Adjective*. List examples of each underneath the words.

- Explain that many words have the same root. Put *danger*, *dangerous* and *endanger* on the board. Elicit the parts of speech and meanings. Ask for some examples. Ask if students know any other words with the same roots.

- Set the task for pairwork.

- Conduct feedback. Ask the whole class to name the parts of speech, then ask different pairs to read their sentences.

Answers

unsafe – *adjective* (opposite)
safety – *noun*
1   unsafe
2   Safety
3   safe

hazard – *noun*
hazardous – *adjective*
1   hazard
2   hazardous

protect – *verb*
protector – *noun* (person or thing)
protection – *noun* (abstract)
protective – *adjective*
1   protective
2   protect
3   protection
4   protectors

dispose – *verb*
disposable – *adjective*
disposal – *noun*
1   disposal
2   dispose
3   Disposable

careless – *adjective*
careful – *adjective* (opposite)
carelessness – *noun*
1   careless
2   careful
3   Carelessness

CLOSURE

- Write one word from each group on the board and ask students to supply the others in the family.

- Review the new vocabulary from the reading text by giving the synonyms to elicit the words.

Lesson 4: Talking about hazards and causes of accidents

Objectives

- to extend vocabulary for safety signs and causes of accidents
- to talk about hazards using the first conditional

Language

- *If* clause structures with imperatives (first conditional)

Vocabulary

- related to safety and accidents

LEAD-IN

- Review key vocabulary, e.g., *hazard, danger, permit, first-aid, emergency* and *shock*. Elicit the parts of speech and any word-building possibilities.

A Discuss meanings of safety signs

- Refer students to the first sign and elicit what it shows. Ask where you might see this sign. Teach or check the meaning of the word *trip*, if necessary.

- Read the exchange in the speech bubbles and ask students to repeat and use this pattern when they do the task.

- Set the task for pairwork and monitor.

- Conduct feedback. Ask individual students what they think the signs show. Ask also for suggestions as to where the signs could be seen, and why.

Answers

1 a trip hazard; be careful of items on the floor
2 an emergency stop button; press when necessary
3 use ear protectors; excessive noise in this area
4 no admittance
5 emergency exit this way
6 wear gloves; dangerous items to handle
7 first-aid; use items from here in case of medical emergency
8 PPE
9 no mobile phones; switch off all mobile phones in this area
10 high voltage; danger of electric shock
11 permits only; only people with permits are allowed in this area
12 danger; forklift trucks

B Understand instructions about what to do in certain situations

- Elicit what the students have to do if the fire alarm sounds: leave the building. Write *IF* on the board and ask students to complete, e.g., *If the fire alarm sounds, leave the building.*

- Clarify the fact that the sentence discusses a possibility rather than a fact.

- Explain that the following sentences also tell us what to do in certain possible situations.

- Set the matching task for individual work and pairwork checking.

- Conduct feedback. Ask the two students in the pairs to start and finish the sentences for the class. Point out that the order of the clauses can be reversed.

Answers

1 If there are cables on the floor, be careful not to trip.
2 If there is a fire, find the nearest emergency exit.
3 If you travel by plane, make sure your mobile phone is turned off.
4 If you go into a restricted area, make sure you have a permit.

C Use prompts to make full sentences

- Draw attention to the use of the *If* clause and the imperative.
- Set the completion task for individual work and pairwork checking. Tell students they will have to include extra words.
- Conduct feedback. Ask individual students to read their sentences to the class.
- An extra task could be to ask students to write the start of three sentences like this, then swap with their partners to finish them. Do one first with the class, e.g., *If the pressure level warning light comes on, ...*
- Conduct feedback. Read the sentences to the class.

Answers

1 If you don't have a permit, don't come in!
2 If someone has an electric shock, get help.
3 Press the emergency stop button if there is a problem with the machine.
4 Watch out for forklift trucks if you go into Area D.

D Write sentences about action to be taken in certain situations

- Set the task for individual work.
- Conduct feedback. Ask students in turn to read full sentences to the class.

CLOSURE

- Review the meanings of the safety signs by asking for words that collocate with *emergency* exit, *ear* protectors, *no* admittance, *high* voltage, *electric* shock, *fork-lift* truck, *mobile* phones.
- Ask students to write two sentences starting with *if* that are true for their own workplaces and read them to the class.

Lesson 5: Talking about risk assessment

Objectives

- to discuss hazards
- to discuss the results of possible actions

Language

- first conditional using modals and imperatives

LEAD-IN

- Review the previous lesson by drawing some safety signs on the board and eliciting what they mean.
- Revise the instructions on what to do in certain situations by giving the first clause and eliciting the imperative, e.g., *If you have an electric shock, ...*

A Identify possible hazards

- Ask students to look at the visual and say what they can see in it.
- Set for pairwork. Students should discuss the possible hazards.
- Conduct feedback. Students present their ideas to the class.

Answers

There is a cable on the floor.
The printer cable is short and could trip someone.
The table by the wall is unsteady.
The water dispenser has leaked water onto the floor.
Someone could trip down the steps.

B 🔊 (CD2 T4) Listen for details to complete the sentences

- Explain that the students are going to listen to Bob's risk assessment of the room and what he wants to do.
- Set the task for individual work and pairwork checking.

- Play the recording for students to listen and complete the task.
- Conduct feedback. Ask individual students to read their sentences in turn.
- Check the meanings of *collapse*, *slip*, *mop up* and *bend.*

Answers

1 If we *put* the printer on table B, it *will* probably collapse.
2 If we *don't move* the cable, someone *will trip* over it.
3 Someone *might slip* if we *don't mop* up the spilt water.
4 If you *move* a heavy object, *bend* your legs before you lift it.

Tapescript
Presenter:
Lesson 5 Talking about risk assessment
B Bob wants to move the printer from table A to table B. Listen to his risk assessment and complete the sentences.

Bob: We're going to have to move things around in here a bit. It's too dangerous at the moment. We need to move the printer, but if we put it on the table over there, it'll probably collapse. And that cable, if we don't move it, someone will trip over it. I know, we'll get rid of that table and move this one over there with the laptop and printer. But first we need to sort out the water dispenser. Someone might slip if we don't mop up the spilt water. And make sure you lift the table carefully. Remember, if you move a heavy object, you must bend your legs before you lift it.

C Look at rules for first conditional sentences

- Write the first sentence from exercise B on the board. Ask students if Bob is definitely going to move the printer or not. Elicit that it is a possibility and the second clause depends on the first. Elicit also that this is not a general rule such as those described with the zero conditional, but is talking about a specific possible situation.

- Tell students that it is a possibility that you will have a holiday in three months' time. If you have a holiday, tell them what you will do. Ask students for possible future situations for them or their families. Construct first conditional sentences from the information.

- Look at the form of the sentences. Elicit and write under a sentence *If* + present tense + , + *will* + base form. Explain that this is the standard form, but it can change slightly.

- Refer students to the rule box. Set the matching task for individual work.

- Conduct feedback. Ask for examples from individual students. Elicit the difference between *will* and *might*, i.e., *If Mohammed takes the exam, he will pass it. If I take the exam, I might pass it!*

Answers

1 If we put the printer on table B, it will probably collapse.
2 Someone might slip if we don't mop up the spilt water.
3 If you move a heavy object, bend your legs before you lift it.
4 If we don't move the cable, someone will trip over it.

D Reorder words to make first conditional sentences

- Remind students that conditional sentences can be written with the *if* clause, or the result clause, first.

- Look at the first example with students and focus on the contraction *will* + *not* = *won't*. Practise pronunciation of this contrasted with *want*.

- Set the task for individual work.

- Conduct feedback. Ask the class to tell you the order while you write the sentences on the board. Ask them to identify the present tenses, future tenses, imperatives and any modals. Explain that we usually put a comma after the *if* clause when it is at the beginning of the sentence.

Answers

2 If the fire alarm sounds, evacuate the building.
3 If you do not follow safety procedures, you could have an accident.
4 If the weather is good, we will complete the job.

E Discuss possible future actions

- Ask students what they will do if someone can't understand their English. Elicit sentences such as *If someone can't understand my English, I will repeat it. I will study harder ... .*

- Ask students to look at the questions and write down one answer for each. Then put them in small groups to tell each other their sentences and discuss them. Monitor and help where necessary.

- Conduct feedback. Ask for a variety of suggestions from the groups.

CLOSURE

- Ask students what they will do if you are not at class next week.

Lesson 6: Talking about past events

Objectives

- to listen to someone talking about incidents in the past
- to review regular and irregular verbs in the past

Language

- past simple

LEAD-IN

- Review the first conditional. Ask questions about actions in the room, e.g., *What will happen if I switch this on?*

A 🔊 (CD2 T5) Listen to identify the problem

- Explain to students that they are going to listen to a conversation between Ahmed and Bob. They identify the problem.
- Play the recording. Ask students to discuss their answers in pairs.
- Conduct feedback. Ask for the answer from the whole group. Ask questions to elicit more details.

Answer

The flow rate was very low.

Tapescript
Presenter:
Lesson 6 Talking about past events
A Listen to Ahmed and Bob talking.
What was the problem?

Ahmed: Had a good day, Bob?
Bob: Not bad, but we had a bit of a panic this morning.
Ahmed: Why? What happened?
Bob: Well, when I got to work at 7, I heard the alarm.
Ahmed: Oh dear, was it serious?
Bob: Well, I checked the control panel of course, and saw the flow rate was really

low. So I opened all the valves to increase the flow rate – but it didn't help much.
Ahmed: So what did you do?
Bob: Oh, I called the pump station and they increased the flow rate. Then it was okay.

B Study the form of the past simple

- Write some short sentences from the listening on the board and ask students to tell you which are the verbs. Underline them and elicit that these are examples of the past tense. Ask what the infinitives of the verbs are. Write them in list form, e.g., *have – had.*
- Divide students into Student 1 and Student 2 pairs. Make sure students are looking at the relevant pages. Students should complete the table individually, then compare answers. Ask students to work in pairs to use their verb tables to answer questions 1 to 5.
- Conduct feedback. Focus on the pronunciation of the endings of the regular past forms.

Answers

come	*came*	adjust	*adjusted*
control	*controlled*	be	*was/were*
hear	*heard*	call	*called*
initiate	*initiated*	increase	*increased*
reach	*reached*	shut down	*shut down*
try	*tried*	stop	*stopped*

1 reach; try; adjust; call
2 initiate; increase; hear
3 control; stop
4 come; shut down; write
5 try: *y* becomes an *i* before adding *ed*

C Use verbs to complete the table

- Set the task for individual work.

- Conduct feedback. Elicit answers from the whole group. Focus on the spelling patterns, i.e., verbs which end in -*e*: add -*d*. Verbs which end in a consonant–vowel–consonant pattern: double the final consonant.

Answers

Verbs that add -*ed*: reach; try; adjust; call
Verbs that add -*d*: initiate; increase; hear
Verbs that double the final consonant: control; stop
Irregular verbs: come; shut down; write

D Find and correct mistakes in a short text

- Set the task for individual work and pairwork checking.

- Ask students to the read correct sentences in turn. Advise students to write down the past form of verbs when they record new vocabulary.

Answers

Bob *came* to work at 7.00 a.m. He *heard* the alarm so he checked the control panel. The flow rate *was* too low. He *tried* to increase the flow rate by opening all the valves, but it didn't solve the problem. He *called* the pump station and they *increased* the pressure.

E Make negative and question forms of the past tense

- Write the first sentence of exercise D on the board, but write 8.00 a.m. instead of 7.00 a.m. to elicit the response *No, he didn't come at 8.00 a.m.* Give more examples of wrong information to correct.

- Ask one student *When did you arrive at college today?* Ask students to repeat your question and put it on the board.

- Refer students to the examples in their books and ask them to choose the correct option.

Answers

1 I *didn't hear* the alarm.
2 What *did* you *do* yesterday?

F Complete a dialogue

- Set the task for individual work and pairwork checking.

- Conduct feedback.

G 🔊 CD2 T6 Listen to check answers

- Play the recording.

Answers

1 turned on 2 checked 3 sounded 4 was
5 didn't initiate 6 adjusted 7 lowered
8 stopped 9 reached

Tapescript
Presenter:
G Listen and check your answers.

Supervisor: What happened yesterday?
Hassan: I turned on the monitoring equipment and checked the system as usual. Then, at 11 o'clock, an alarm sounded.
Supervisor: What was the matter?
Hassan: The flow rate was too high.
Supervisor: Did you initiate an emergency shutdown?
Hassan: No, I didn't initiate an emergency shutdown. I adjusted the flow rate and lowered the pressure in the pipeline. The alarm stopped when the flow rate reached normal levels.

CLOSURE

- Ask students to ask/tell each other about any situations they have been in when an alarm sounded.

Lesson 7: Talking about actions in progress in the past

Objectives

- to contrast and understand the difference between past simple and past continuous
- to practise using the past continuous

Language

- past continuous

LEAD-IN

- Review the past simple by giving a quick oral test on regular and irregular past verb forms. Then elicit negatives of different statements and also how to make questions from them.

A Establish the concept of past continuous

- Ask students what they are doing now to elicit the present continuous for an action in progress. Transfer this to the past by looking at your watch and telling them what you were doing at this time yesterday. Elicit what they were doing at this time yesterday. Focus on the fact that the action/activity was ongoing and not finished at this precise time.

- Put some contrasting sentences on the board. List what you did this morning, e.g., *I got up at 7.30. I had a shower. I had my breakfast at 8.00 a.m. I drove to work at 8.30.* Ask for some similar lists from students. Then focus on one activity, e.g., *I was having breakfast when the phone rang.* The breakfast was in progress when it was interrupted by the phone.

- Set for pairwork. Refer students to the two pictures. Ask them to choose the correct sentences.

Answers

2   At 8.00 p.m. she drove home.
    (The action was not in progress at 8.00 p.m.)
2   They were leaving the workshop when the phone rang.
    (The action of leaving was in progress when it was interrupted by the phone ringing.)

B Study the past continuous

- Set the reading and completion task for individual work and pairwork checking.

- Conduct feedback.

- Highlight how we form the past continuous: past tense of *be* + verb + *-ing*. Practise pronunciation of weak forms *was* and *were* when using the past continuous.

- Tell students some more things you did yesterday and write them on the board, e.g., *I drove to work. I cooked dinner. I watched television.* Ask students to work in pairs to make a past continuous sentence from each of these. Ask some students to read their sentences to the class.

Answers

At 5.00 p.m. we *were working* in the workshop.
I *was checking* the pressure when the alarm sounded.

C Practise using the past continuous

- Set for pairwork. Ask students to ask and answer the questions using the past continuous. Point out that they may also need to use the past simple for actions that were not continuous and/or state verbs.

- Conduct feedback. Bring the questions into the whole group.

Example answers

1   I was working in the office.
2   I was studying electronics.
3   Mehdi was sitting in class. He was doing his homework. Two teachers were talking in the corridor.
4   The economy was getting better.
5   Yes, I have.
6   I was painting the ceiling when I fell off the ladder and I broke my leg.

CLOSURE

- Set for pairwork. Ask students to write the first parts of five past continuous sentences for their partners to finish. They should read them to the class. Do one with them first, e.g., *I was marking your homework when … .*

Lesson 8: Reporting incidents

Objectives

* to learn vocabulary for and read about reporting incidents
* to write report cards using past simple and continuous tenses

Vocabulary

* words related to reporting incidents

LEAD-IN

* Briefly review the past continuous by asking what was happening when students arrived at class today.

A Check vocabulary related to reporting incidents

* Discuss with students what procedures have to be followed when there is an incident at work.
* Ask them if they have any systems in place to help improve safety in their workplaces.
* Explain they are going to look at some words often used when reporting incidents.
* Set the task for individual work.
* Conduct feedback. Elicit answers from the full group. Practise pronunciation, particularly word stress.

Answers

1   **a**   to watch carefully
2   **b**   to persuade
3   **a**   in charge of
4   **b**   to make small changes
5   **b**   to get rid of
6   **a**   rules for how to do things

B Read and complete a text

* Ask students if they have heard of the STOP system. If so, can they say what it is?

* Set the task for individual work and pairwork checking.
* Conduct feedback. Ask different students to read consecutive sentences.

Answers

The STOP system is based on the following idea. Everyone is *responsible* for safety, and all injuries can be eliminated if safe *procedures* are reinforced. STOP report cards can be positive or negative because people who work safely should be told, as well as those who do not. The key to reducing incidents and injuries is to *observe* people, talking with them to *encourage* safe work practices and, therefore, *modify* their behaviour to *eliminate* unsafe acts and behaviours.

C Answer questions about the text

* Set the task for individual work. Students write their answers, then read them to the class.
* Ask students if they think this is a good system.

Answers

1   Everyone.
2   To encourage safe work practice and modify their behaviour.
3   Injuries and incidents can be eliminated.

D 🔊 (CD2 T7) Complete STOP cards with correct verb forms

- Explain that students are going to read three STOP report cards. They should put the verbs in brackets in the correct form to complete the texts. Remind them to think about whether the actions were in progress or interrupted at a particular time in the past.

- Conduct feedback.

- Ask why the verbs are in the different forms. Point out that we often use the past continuous at the beginning of an account to describe background events.

- Play the recording for students to listen and check their answers.

Answers

1
1   noticed
2   bent
3   put
4   congratulated

2
1   was washing
2   noticed
3   was
4   informed
5   repaired

3
1   was malfunctioning
2   was flashing
3   thought
4   was
5   notified
6   sent

Tapescript
Presenter:
Lesson 8 Reporting incidents
D Complete the STOP report cards with the verbs in the correct tense. Then listen and check.

Voice 1:   A technician noticed a piece of wood with nails sticking out in front of the equipment lockup. First, he bent the nails over with a hammer, then put the wood in the garbage container. I congratulated him on his action.

Voice 2:   While I was washing my hands, I noticed a broken mirror frame in the bathroom, which was in danger of falling. I informed Administration, and the company repaired the frame.

Voice 3:   The light in a classroom was malfunctioning before lunch. The light was flashing on and off. I thought it was a short circuit and a fire hazard. I notified Administration and they immediately sent someone to repair the light.

E Write a STOP card report

- Focus attention on the pictures and elicit key words such as *cable*, *trip*, *heavy object* and *lift*.

- Set for pairwork. Ask them to choose one of the pictures and discuss the incident and what they should write in a report card. Students write the STOP card report individually and check in pairs afterwards.

- Ask some students to read their reports to the class.

CLOSURE

- Review vocabulary by giving definitions to elicit the words.

- Ask students if they can tell you about an incident that happened recently at their workplace.

Lesson 9: Asking about incidents

Objective

- to review and practise using question words when asking about incidents in the past

Language

- question words
- past tense questions

LEAD-IN

- Review vocabulary related to reporting incidents. Ask if there have been any incidents at work since the last class.

A Choose appropriate question words

- Ask students how many question words they know and list them on the board. Do not add any at this stage.
- Refer students to the table. Set the task for pairwork.
- Conduct feedback. Ask for the question word, then an example of what it can refer to. Ask the group for an example question using each word.
- In pairs, each student gives the other the question word(s) to start a question.

Their partner finishes it. Do one first with students, e.g., *How much ...? How much did your car cost?*

Answers

See table below.

B Complete questions in a dialogue

- Go over the word order in questions. Give the following example: *Who installed it?* See if students can restate it in the passive voice, i.e., *Who was it installed by?* Remind students to look carefully at the answers while doing this exercise to decide whether they need to use the passive or active voice.
- Set the task for individual work and pairwork checking.
- Conduct feedback.

Question word	Example	Use
What ...?	*What's that?*	things
Where ...?	*Where do you live?*	place
When ...?	*When do you get up?*	time
Who ...?	*Who did you meet yesterday?*	people
Why ...?	*Why are you learning English?*	reason
Which ...?	*Which car do you prefer?*	choice
Whose ...?	*Whose dictionary is that?*	ownership
How ...?	*How do you make a cake?*	method
How often ...?	*How often do you go to the gym?*	frequency
How much ...?	*How much time have you got?*	quantity
How old ...?	*How old is your son?*	age
How long ...?	*How long did you wait?*	length of time
How far ...?	*How far is your home from here?*	distance

Answers

Where is it?
When was it installed?
Who *was it installed by?*
Why was it installed?
How *did he install it?*
How long did it take to replace?

C 🔊 (CD2 T8) **Listen to a description of an incident**

- Set the task for pairwork. Check the meaning of *wrapped, handkerchief, medical block, stitches* and *fainted.*

- Play the recording for students to listen.

- Check understanding before students move onto the next part of the task. Ask *What injury did Bob have? How did he do this? Where did he go?*

- Students complete the task in pairs.

- Conduct feedback. Go through their questions with the whole group.

Tapescript
Presenter:
Lesson 9 Asking about incidents
C **Listen to Bob describing an incident. Think of questions you would like to ask about it.**

Bob: I cut my hand when I was working. I was changing a valve, and while I was removing it, my pipe wrench slipped and cut into my hand. I wrapped my hand in my handkerchief and got someone to take me to the medical block. The doctor said I needed stitches, and I thought 'Oh no!' Then, when he was stitching up my hand, I fainted!

D 🔊 (CD2 T9) **Listen to questions**

- Play the recording for students to listen and compare their questions. Elicit the questions Ahmed asked by giving prompts if necessary.

- Elicit the fact that the second question uses the past continuous tense. *Why were you changing the valve?* Highlight the form, drawing attention to the use of the auxiliary *were* and the use of *-ing.*

Answers

When did it happen?
Why were you changing the valve?
Who took you to the medical block?
Did it hurt?
What did the doctor do when you fainted?

Tapescript
Presenter:
D **Listen to Ahmed asking Bob about the incident. Were his questions the same as yours?**

Ahmed: When did it happen?
Bob: Yesterday evening, just before the end of my shift.
Ahmed: Why were you changing the valve?
Bob: Oh, it was old and faulty.
Ahmed: Who took you to the medical block?
Bob: One of the roustabouts.
Ahmed: Did it hurt?
Bob: Not at first. It started hurting when I had the stitches, and it's quite sore now.
Ahmed: What did the doctor do when you fainted?
Bob: I don't know! When I woke up, I was lying on the couch!

E **Role-play the situation**

- Set the task for pairwork. Monitor and help where necessary.

- Ask some pairs to repeat their dialogues for the class.

- Ask students to write down the dialogues from memory.

CLOSURE

- Give some answers and elicit the questions (or simply question words) from the students, e.g., *My house is five kilometres away.*

Lesson 10: Talking about the golden rules

Objectives

- to develop intensive reading skills
- to learn vocabulary connected with different operations
- to practise speaking about safety in different areas

Vocabulary

- words related to safety rules

LEAD-IN

- Elicit what students know about the BP (British Petroleum) company.

A Discuss safety rules

- Tell students that in this lesson they will be looking at BP's 8 golden safety rules. First, they should discuss in small groups what they think are important safety rules for people working in the oil, gas and petrochemical industries and note them down.

- Conduct feedback. Discuss some of the students' ideas with the whole group.

B Find synonyms in a text

- Ask students to read the introduction to the golden rules. Ask comprehension questions to ensure general understanding, e.g., *What does the policy state? What happens before work is started? What must be worn? When must people stop work?*

- Set the task to find words and phrases in the text for individual work and pairwork checking.

- Conduct feedback. Elicit answers with the whole group. Deal with any other unfamiliar words such as *conduct*, *potential*, *obligation* and *scenario*.

Answers

1 harm
2 enforced
3 risk assessment
4 competent
5 site requirements
6 emergency response plans
7 commencement
8 obligation

C Read a text to match rules and explanations

- Refer students to the word box. Point out that this is formal language, e.g., in general spoken English we would replace *confined space entry* with *going into small spaces*.

- Set the task for individual reading. Advise students to scan the rules to look for key words that they associate with them. Do the first one together. The word *permit* is in line 3. Reassure students that they do not need to understand every word to have a general understanding of the text.

- Conduct feedback.

- Allow students to choose five words from the text that are unfamiliar to them and check the meanings in a dictionary. Pool their findings with the whole group.

Answers

1 Permit to work
2 Working at heights
3 Energy isolation
4 Confined space entry
5 Lifting operations
6 Management of change (MOC)
7 Driving safety
8 Ground disturbance

D Discuss the rules

- In their original groups, students compare the rules they discussed at the beginning of the lesson with BP's golden rules. In these groups, they discuss which rules are most important and relevant to them. They must be prepared to say why to the whole group.

- Discuss with the whole group.

CLOSURE

- Review new vocabulary from the lesson.

Answers

A

Example answers

If someone walks near the forklift, the box will fall on them.
If someone walks near the work bench, the saw will fall and cut them.
If someone trips over the power cord, the drill will fall and hurt them.

B C

Students' own answers.

D

Student 1's answers:

Date and time of incident	June 2003, 2.00 p.m.
Where the incident took place	In a workshop.
Description of incident	Fall from a ladder. Electrician suffered severe head injuries.
Action taken	Driver taken to hospital.

Student 2's answers:

Date and time of incident	22nd September, 2005, 4.00 p.m.
Where the incident took place	On a steep hill.
Description of incident	Vehicle crash. (The vehicle should only have had one person – it had two. Safety head gear and ear protection were not worn. Brakes were not functional and brake fluid was low.)
Action taken	Driver taken to hospital.

Assess your skills: Talking about safety

• Refer students to the self-assessment grids.

Word list

aisle
ankle
barrier
bone
buoy
chemical
chest
commencement
competent
cone
confined
contusion
dislocation
disposable
disposal
disturbance
ear protectors
elbow
eliminate
encourage
face mask
finger
fire extinguisher
first-aid
flare
fracture
handle
harm
harness
hazard
hazardous
heavy duty
injure

injury
isolation
joint
knee
laceration
lifeline
lightweight
limb
modify
multiple
neck
observe
permit
policy
potential
procedure
prohibit
protect
reinforce
resistant
responsible
risk
scenario
severe
shoulder
sprain
superficial
thumb
toe
torso
trip
twist
wrist

UNIT 7

MAKING COMPARISONS

Lesson 1: Making general comparisons between two things

Objective

- to review the use of comparative adjectives

Language

- comparative form of adjectives

Vocabulary

- adjectives

LEAD-IN

- Ask what the temperature is today and write it on the board. Write temperatures of other cities/countries on the board. Elicit which are hotter and which are colder than here.

- Elicit comparisons between pairs of objects in the classroom and/or pairs of cities on the board.

A Establish the concept of comparing things and people

- Ask students to read through the *Comparative adjectives* information box and add a few examples of things you have already compared on the board.

- Tell students that you have two books. One is interesting, but the other is more interesting. Elicit other adjectives that make comparatives in this way. Highlight the different comparatives for long and short adjectives.

- Refer students to the tables and the word box. Set the task for pairwork.

- Conduct feedback. Name the adjectives and ask for the comparatives from the whole group.

Answers

Short adjectives:

One-syllable	Ending in CVC	Ending in -*e*
fast – faster long – longer	big – bigger hot – hotter	safe – safer

Ending in -*y*	Irregular
heavy – heavier easy – easier dirty – dirtier	bad – worse good – better

Long adjectives:

Two- or three-syllable
useful – more useful flexible – more flexible expensive – more expensive dangerous – more dangerous complex – more complex

B Write sentences using comparative adjectives

- Set the task for individual work and pairwork checking.

- Conduct feedback. Ask different students to read a sentence each.

- If more practice is necessary, ask students to write three more gapped sentences like these to exchange with their partners.

- Conduct feedback. Give examples to the class.

Example answers

1 faster than
2 more practical/expensive than
3 more flexible
4 more dangerous than
5 heavier than
6 longer than
7 more complex than

C Discuss different methods of recovery

- Direct attention to the visuals. Elicit what a nodding donkey is (or ask students to check in the glossary).
- Set the task for group work. Encourage students to use comparatives in their discussion.
- Conduct feedback. Elicit ideas.

D ◐ (CD2 T10) Listen to identify different words used

- Read through the list of words in the word box with students. Elicit which they think they will hear in the different descriptions.
- Play the recording for students to listen and circle the words used. Students compare answers.
- Conduct feedback. Ask which words were used and in which description, nodding donkey or offshore platform.

Answers

Nodding donkey: small; cheap to run; easy to maintain; traditional; reliable
Offshore platform: sophisticated; dangerous; effective; useful; drills deeply

Tapescript
Presenter:
Unit 7 Making comparisons
Lesson 1 Making general comparisons between two things

D Listen and circle the words and phrases that are used to compare the different types of oil recovery.

Voice: Nodding donkeys have been around for a long time. They are a traditional way of extracting oil. They are smaller than offshore platforms, and cheaper to run. They are also clearly easier to maintain than offshore platforms. On the other hand, they cannot be used offshore, so offshore platforms were needed when oil started to be extracted out at sea. Offshore platforms are designed to drill more deeply than the traditional nodding donkey. They are more sophisticated and effective, but consequently, they are potentially more dangerous. A nodding donkey has the advantage of usually being more reliable than an offshore platform, but an offshore platform is ultimately more useful.

E Write sentences comparing methods of oil recovery

- Ask students to write six sentences comparing the methods. They should use the words in the box as prompts. Students compare their sentences with a partner.
- Conduct feedback. Ask different students to read one of their sentences to the class.
- For extra practice, ask students in pairs to compare two types of vehicle/job or equipment. They should write down three or four sentences, naming the items A and B.
- Ask different pairs to read their sentences to the class. The other students have to guess which things are being compared.

CLOSURE

- Give students an adjective and they have to make a comparative sentence using this adjective. Continue with other adjectives and try to give both long and short.

Lesson 2: Making more specific comparisons

Objective

- to describe greater and smaller differences between things

Language

- adverbs of degree: *much* and *slightly*

Vocabulary

- drill system parts
- casing adjectives

LEAD-IN

- Review comparative forms from the previous lesson by giving adjectives and eliciting comparative forms. Ask students to spell the comparatives for you to write on the board.

A Identify and label a diagram

- Ask students to look at the diagram and identify it. (It is a drill in an oil well.) Check the meaning of *hole*, *drill*, *bit*, *string*, *casing* and *pressure*.
- Set the labelling task for pairwork.

B ◉ (CD2 T11) Listen to check answers

- Play the recording for students to listen and check answers. Note that *drill string* is not mentioned on the recording.
- Set for pairwork. Ask students to discuss the diagram. Make comparisons using words and information from the recording.
- Elicit the opposites of *deep*, *strong*, *wide*, *thick*, *low*.

Answers

1 surface 2 drill string 3 casing 5 drill bit

Tapescript

Presenter:
Lesson 2 Making more specific comparisons

B Listen and check your answers.

Voice: This diagram shows an oil well casing. Casings are used to protect the weaker, upper parts of the oil well, and stop the sides of the well falling in. This is particularly the case with deep formations, as they have higher pressures than shallow formations. When a well is drilled, the top diameters are always larger than those deeper in the well. Consequently, the hole and casing diameter are narrower at the bottom of the well than at the top. The hole is always slightly bigger than the casing, so that a cement bond can be pumped between the outside of the casing and the wall of the hole. After the first section of a well is drilled, a wide diameter casing is fitted inside the hole. A drill bit smaller than the casing is then used to drill the next section of the hole.

C Correct mistakes in comparative forms

- Set the reading and correcting task for individual work and pairwork checking. Remind students of the spelling rules in the previous lesson.

- Conduct feedback. Ask different students to identify the mistakes they corrected and spell corrections.

Answers

1   … have *higher* pressures … to protect *weaker* upper formations …
2   … are *wider* than …
3   … is *narrower* at the bottom …
4   … slightly *bigger* than …
5   A drill bit *thinner* than …

D Using degrees of comparison

- Ask students if they can remember whether in the description the hole is a lot bigger than the casing. Elicit *slightly*. Elicit *much* by indicating various objects in the room and comparing their size, e.g., *A is much bigger than B*. Practise both by discussing cars, temperatures, etc.
- Refer students to the *Adverbs of degree* information box to check their understanding.
- Ask students to look at the visuals. Set the task for individual work and pairwork checking.
- Conduct feedback.

Example answers

1   A is much *smaller* than B.
2   A *is much smaller than* C.
3   B *is bigger than* A.
4   C *is much bigger than* A.
5   B *is slightly bigger than* C.

E 🔊 (CD2 T12) Describe liquids

- Refer students to the visuals. Set the task for pairwork. Ask them to think of adjectives to describe each visual. Write the adjectives they have thought of on the board.
- Play the recording for students to listen and note down two adjectives for each liquid.
- Pairs compare their adjectives.
- Conduct feedback. Add the adjectives to those on the board.

Answers

Water: transparent; cheap; essential; inert
Crude oil: viscous; heavy; flammable
LPG: combustible; dangerous; explosive
Acid: corrosive; dangerous

Tapescript
Presenter:
E **Listen and write down two adjectives that are used to describe each liquid, e.g., *viscous, heavy*, etc.**

Voice 1: Crude oil is viscous, much more viscous than water. It's also heavy, isn't it?
Voice 2: It's heavier than water … and oil is flammable. LPG gas is highly combustible, so it's quite dangerous. It's more explosive than oil, for instance.
Voice 1: What about acid? Acid is corrosive. It's also dangerous.
Voice 2: Yes, I think acid is slightly more dangerous than oil or gas.
Voice 1: Water is generally cheap. But it's important if you don't have it – it's essential for life.
Voice 2: Chemically, water is transparent and inert.

F Compare different liquids

- Set the task for pairwork. One person chooses two of the liquids; the other person describes them, using a comparative structure.
- Conduct feedback. Ask for different comparisons.
- Extend the activity by asking students to think of other liquids they can compare.

CLOSURE

- Write pairs of words on the board. Elicit comparisons using *slightly* and *much*: tall buildings, machines, busy places, interesting magazines/newspapers, exciting films, etc.

Lesson 3: Comparing more than two things

Objective

• to compare more than two things

Language

• superlative form of adjectives
• use of *more* and *most*; *less* and *least*

Vocabulary

• adjectives to describe materials

LEAD-IN

• Tell students that yesterday you were busy. Elicit who in the class had a busier day. Find a third student whose day was even busier. Elicit *busiest*. Write the three forms on the board.

A Discuss qualities of materials

• Refer students to the table. Check the meanings of the adjectives. Elicit what the difference between one tick and two ticks is (one tick for *hard* and two ticks for *very hard*).

• Set the task for pairwork.

• Conduct feedback. Discuss students' suggestions for the rest of the table. Do not confirm whether they are right at this point.

B 🔊 (CD2 T13) Listen to check answers

• Set the task for individual work and pairwork checking. Play the recording for students to listen and complete the table.

• Conduct feedback.

Example answers

See table below.

Tapescript
Presenter:
Lesson 3 Comparing more than two things
B Listen and complete the table above.

Voice: As regards hardness, rubber is not hard at all. Gold and glass can be classed as hard. Wood is harder than rubber, but steel is the hardest substance. If we compare price, it depends what you're making. In production terms, steel is more expensive than wood, but gold is the most expensive. Rubber is the most elastic substance, although you might be surprised to hear gold is also elastic. Rubber and wood are both combustible substances. Rubber is more combustible than glass, gold or steel, but wood is the most combustible. Wood is probably the

	Hard	Expensive	Elastic	Combustible	Versatile
glass	✓	✗	✗	✗	✓
rubber	✗	✗	✓✓	✓	✗
wood	✓	✗	✗	✓✓	✓✓
steel	✓✓	✓	✗	✗	✓
gold	✓	✓✓	✓	✗	✓

most versatile material, but steel is also quite versatile. Glass, rubber and gold are used to make a narrower range of products because they are less versatile. Perhaps rubber is the least versatile.

C Use comparative forms to complete sentences from a recording

- Set the task for individual work.
- Conduct feedback.
- If students need extra practice, ask them to work in pairs to list different materials and write some sentences comparing them.

Example answers

1 Wood is *harder than* rubber, but steel is *the hardest* substance.
2 Steel is *more expensive* than wood, but gold is *the most expensive.*
3 Rubber is *the most elastic substance.*
4 Wood is *the most versatile substance.*

D Check rules for comparative and superlative forms

- Ask students what we call the third form: *superlative.* Elicit when we use -*est* and when we use *more.*
- Read through the *Superlative adjectives* information box and ask students to complete the rule.
- For extra practice, students can give their partners an adjective to put into the superlative form. Alternatively, write a list of long and short adjectives on the board. Elicit the superlatives from the whole class.

Answers

Forming superlatives with one-syllable adjectives: *the* + adjective + -*est*
Forming superlatives with longer adjectives: *the* + *most* + adjective

E Compare different engines using superlative forms

- Refer students to the visuals. Elicit the following: *jet engine, internal combustion engine, steam engine* and *electric motor.*
- Read through the question prompts with students. Check meanings of unfamiliar vocabulary or concepts such as *sophisticated* and *environmentally friendly.*
- Discuss the first question with the whole class. Emphasize that you are looking for their opinions rather than correct answers.
- Set the task for pairwork.
- Conduct feedback. Elicit ideas for group discussion. Encourage students to give their opinions using *In my opinion/I agree with you*, etc.

F Use *less* and *the least* in making comparisons

- Use one or two sentences from exercise E to elicit the concept of *less* and *least*, e.g., *Steam engines are less environmentally friendly than electric motors. Electric motors are the least dangerous.*
- Set the task for pairwork. Students complete the table with their own ideas.
- Conduct feedback.

CLOSURE

- Initiate a short discussion about what students find easier/easiest/more difficult/most difficult about learning/using English.
- Finish by saying *Using comparative and superlative adjectives is easier now than it was a few weeks ago!*

Lesson 4: Comparing metals

Objectives

- to learn vocabulary for qualities of different metals
- to discuss the qualities of different metals

Language

- consolidation of comparative structures
- word-building: nouns and adjectives

Vocabulary

- adjectives and nouns to describe qualities of metals

LEAD-IN

- Elicit names of metals. Write them on the board. Ask students for one adjective to describe each metal.

A Choose adjectives to describe metals

- Read the first sentence with students. Elicit whether this is true or false.
- Set the task for individual work and pairwork checking. Encourage students to guess the answers they do not know.
- Conduct feedback. Elicit the meaning of the underlined words.

Answers

1 T 2 T 3 F 4 F 5 T 6 T 7 T 8 T

B Make nouns from adjectives

- Set the task for individual work and pairwork checking.
- Conduct feedback. Ask students to give the words and spell them for you to write on the board. Elicit any other nouns students know with these endings.

Answers

1 durability – durable 2 ductility – ductile
3 conductivity – conductive 4 fusibility –
fusible 5 hardness – hard 6 brittleness –
brittle 7 malleability – malleable 8 lustre –
lustrous

C 🔊 (CD2 T14) Practise syllable stress

- Ask students to read through the nouns in exercise B and underline where they think the stress is. Do the first one together with students.
- Compare answers, then play the recording for students to check.

Answers

1 durability 2 ductility 3 conductivity
4 fusibility 5 hardness 6 brittleness
7 malleability 8 lustre

Tapescript
Presenter:
Lesson 4 Comparing metals
🅲 **Listen and underline the stressed syllable in each noun in exercise B.**

Voice: 1 durability

2 ductility
3 conductivity
4 fusibility
5 hardness
6 brittleness
7 malleability
8 lustre

D Match qualities and definitions

- Set the task for individual work and pairwork checking.
- Conduct feedback. Ask for nouns from individual students.

Answers

2 fusibility 3 lustre 4 elasticity
5 malleability 6 ductility 7 brittleness
8 conductivity 9 durability

E Compare qualities of different metals

- Elicit the meaning of the three ticks. Set for individual work and pairwork checking. Students will need to use *less* and *the least* as well as comparatives and superlatives to complete the sentences.
- Conduct feedback.

Answers

1 harder than 2 the hardest 3 a higher/the highest 4 less thermal conductivity than 5 the least malleable

F Discuss properties

- Set the task for groupwork. Ask one person in each group to be a secretary and note down the group's ideas.
- Conduct feedback. Elicit ideas from groups. List suggestions on the board.

G ● (CD2 T15) Listen to check answers

- Play the recording for students to listen and see if the ideas on the recording are the same as the ones on the board.
- Conduct feedback.

Answers

Chains: must have tensile strength, hardness, be malleable, resist chemical corrosion, have durability. Must not be brittle or ductile.
Storage tanks: must be durable, not too brittle, have hardness, resist corrosion. Hot water tanks must not be conductive or fusible. Don't need to have lustre.
Heating elements: need conductivity, to be malleable, need durability, must be hard.

Tapescript
Presenter:

G Listen to the conversation and check your answers.

Voice 1: Chains need to have tensile strength and hardness. They should be malleable in order to bend the metal into links.
Voice 2: Yes. They also need to resist chemical corrosion and have durability, especially if they are in the open air, on the deck of a rig, or on a boat. They shouldn't break or change their length, so they can't be brittle or ductile.
Voice 1: Storage tanks are durable and can't be too brittle. They have to have hardness and resist surface indentations and corrosion.
Voice 2: What about hot water tanks?
Voice 1: Well, they shouldn't be conductive or fusible. They don't need to have lustre.
Voice 2: The metal for heating elements needs conductivity to heat the air or water around the element. They need to be made of a malleable material to be bent into the correct shape.
Voice 1: Yes, and they also need durability and their surface needs to be hard so that they remain smooth.

CLOSURE

- Ask students to discuss the metals found in their workplaces. They should say what qualities these metals have and why they are used.

Lesson 5: Talking about pigging

Objectives

- to read about pigs and pigging
- to practise skimming and scanning skills
- to discuss pigs and the pigging process

Vocabulary

- words related to cleaning and pigging

LEAD-IN

- Write the word *pig* on the board. Elicit what the students know about this.

A Study vocabulary related to pigging

- Set the task for individual work and pairwork checking.
- Conduct feedback. Ask students which word is the odd one out, and why.
- Check they understand the meaning of the other words.
- Can they guess how these words relate to pigging?

Answers

1   seal (a stopper or fastener; the others are types of material)
2   plug (something that fills a hole; the others all describe liquid)
3   by-pass (a verb or adjective describing something that passes around something else; the others are all nouns)
4   debris (remains or residue; the others are all adjectives)

B Answer questions about a pig

- Refer students to the diagram. Check any problem vocabulary and ask them to answer the questions in pairs.

- Conduct feedback. Elicit ideas, but do not confirm whether they are right at this point.

C Read texts to confirm answers

- Refer students to the texts. Before they read them, elicit where the texts might be from (a dictionary, a glossary and an advertisement).
- Set the reading and checking the task for individual work. Encourage students to skim through the texts quickly.
- Conduct feedback.
- Write these prompts on the board and ask students for the contexts: *in order to, propelled down, work up to, most economical.*

Answers

1   It inspects, cleans and/or gives information about the pipeline.
2   No.
3   Foam, but also plastic, rubber and steel.
4   Both.

D Answer *true/false* questions about pigging

- Set the task for individual work and pairwork checking. Encourage students to look at the *true/false* statements before they read the text and to scan to find the answers.

- Conduct feedback. Ask whether the statements are true or false. Students correct the false statements.

Answers

1 F
2 F
3 T
4 F
5 F
6 T
7 T

E Discuss problems related to the pigging process

- Set the task for pairwork.
- Ask students to discuss what can go wrong with the pigging process, using some of the words given. If students are unclear about the process, start them off by discussing what is likely to happen if there is a lot of corrosion in a pipe as a class.
- Conduct feedback. Students give their comments to the full group. Ask what the other words in the box referred to in the text.

CLOSURE

- Ask students to finish these words and put them in context: *solu...*, *fric...*, *lubric...*, *elev...* and *deb...* .
- Ask students to describe the different uses of a pig.
- Discuss how often students come into contact with the pigging process, and which types.

Lesson 6: Expressing similarities

Objective

- to practise describing similarities and differences between objects

Language

- comparative structures

LEAD-IN

- Establish the context by looking at temperature or something similar, e.g., *Yesterday it was hot. Today is hot. Today is as hot as it was yesterday.*

A Describe similar objects

- Refer students to the visuals. Elicit some comparisons.
- Set the task for individual work and pairwork checking.
- Conduct feedback.

Answers

1	T
2	F
3	F
4	F
5	T
6	T

B Describe differences and similarities

- Set the task for pairwork.
- Conduct feedback.
- Write the structures on the board: *... is as ... as, is not as ... as, both ... are ..., they are both ...* .
- For extra practice, ask students in pairs to draw two similar diagrams of objects that are similar, but not identical. The pairs swap diagrams and discuss or write sentences to describe the similarities and differences.

Answers

Sentences 1, 3 and 6 focus on the similarities.
Sentences 2, 4 and 5 focus on the differences.

C Write sentences to compare two objects

- Set the task for pairwork, then individual work. Ask students to discuss the comparisons first in pairs, then write the sentences individually.
- Conduct feedback. Ask students to read their sentences in turn to the class.

Example answers

1	C is bigger than D.
2	Both C and D have a light colour.
3	They are both the same shape.
4	C and D are both water tanks.
5	D is not as expensive as C.
6	They are both made of plastic.

D Choose correct alternatives to complete sentences

- Set the task for individual work.
- Conduct feedback.
- Ask students to make similar comparisons between things they can see now.

Answers

1 dense
2 than
3 as
4 Both
5 are both

E Write a description of two pigs

- Refer students to the two pigs. Elicit initial comparisons.
- Set the writing task for individual work.
- Conduct feedback. Ask some students to read their comparisons to the class.

CLOSURE

- Ask students to think of items they often use at work or at home and to describe the differences between them, e.g., *The coffeemaker at home is not as big as the coffeemaker at work. The cheese sandwiches in the canteen are not as tasty as the chicken sandwiches at home. Both my computer at work and my computer at home are very fast.*

Lesson 7: Measuring temperature

Objectives

- to practise listening to and making comparisons between types of thermometer
- to write descriptions of different thermometers

Language

- comparative structures

Vocabulary

- words related to thermometers

LEAD-IN

- Books closed. Ask students what we use to measure temperature. How many different types of thermometers can they name?

A Discuss thermometers

- Books open. Refer students to the visuals and set the discussion task for pairwork.
- Conduct feedback. Elicit ideas. Put the words *industrial*, *medicinal* and *domestic* on the board.
- Ask for specific examples of when thermometers are used in these different contexts.

B ◉ (CD2 T16) Listen to complete a table with names of liquids used in thermometers

- Elicit types of liquids contained in thermometers.
- Check suggestions against the instructions for the task.
- Set the listening task for individual work and pairwork checking.
- Play the recording for students to listen and complete the table.

Answers

1 Alcohol
2 Mercury
3 Pentane

Tapescript
Presenter:
Lesson 7 Measuring temperature
B Some thermometers contain mercury, some contain pentane and some contain alcohol. Listen to the comparison of thermometers and complete the table with the names of the three liquids.

Voice: Thermometers have many industrial, medicinal and domestic uses. A good device is accurate to nought point one degrees Celsius. Typical liquids used in thermometers are mercury, alcohol and pentane. Mercury has the widest range and is particularly good for measuring high temperatures up to five hundred and ten degrees Celsius. Alcohol thermometers have the shortest range, but they can be used to measure temperatures between minus eighty degrees and seventy degrees Celsius. For very low temperatures however, pentane is used, as it can measure temperatures as low as minus two hundred degrees.

C Complete a text with correct comparative and superlative forms

- Review comparative and superlative forms. Set the task for individual work. Ask students to complete the first paragraph, then check answers.

- Ask students to complete the task, then compare answers.

- Conduct feedback. Ask students to read sentences in turn. If necessary, explain why a comparative or superlative form is used in the different cases.

- For more practice, ask students to cover the text and reread it, making some mistakes for students to correct, e.g., *alcohol has a longer range ... .*

Answers

1 shortest
2 longest
3 lower
4 the lowest
5 higher
6 weaker
7 stronger than
8 more popular than
9 longer
10 more quickly than

D Discuss different types of thermometers

- Refer students to the visuals. Set the discussion task for pairwork.

- Conduct feedback. Elicit suggestions.

E Write a comparison between two thermometers

- Set the writing task for individual work.

- Ask some students to read their comparisons to the class.

CLOSURE

- Books closed. Ask students to name the different types of thermometer, what they contain and what they are used for. If students have difficulties comparing the thermometers in the drawings, ask them to choose other thermometers to compare (they could look on the Internet).

Lesson 8: Describing states of matter

Objectives

- to use vocabulary to talk about states of matter
- to listen and complete notes
- to review numbers

Vocabulary

- words related to chemistry and states of matter

LEAD-IN

- Write the word *matter* on the board. Elicit the three states with some examples for each.

A Discuss states of matter

- Ask students to identify the three states in the visuals as a whole group. Then set the discussion task for pairwork.
- Conduct feedback. Elicit comments and ideas.
- Check understanding by eliciting examples of solids, liquids and gases.

Answers

solid; liquid; gas

B Identify solids, liquids and gases

- Refer students to the word box. Set the identification task for pairwork.
- Conduct feedback. Ask students to write down three more items whose states are not so clear to ask the rest of the class.

Answers

Gas: water vapour; LPG
Liquids: solvents; mercury; solutes; crude oil
Solid: ice; sugar; grease

C ◆ (CD2 T17) Listen to a mini-lecture to complete notes

- Explain that students are going to listen to a mini-lecture about matter. Before they listen, ask them if they can explain the difference between a solid, a liquid and a gas.
- Play the recording. Ask students to listen to see how the different states are discussed.
- Give students time to read the notes. Play the recording again for them to complete the notes. Emphasize that notes do not have to form full sentences. Students do not need to worry about every word being the same as on the recording.
- If students find this difficult, pause the recording after each section. Let students compare their answers with a partner.
- Conduct feedback. Ask for sentences from different pairs.
- Finish by asking some students to read the completed paragraphs to the class.

Answers

Definitions:
definite shape; weaker than the forces in solids; have no definite volume or shape; weaker than liquids

Changing states:
giving molecules more energy; liquid; changes it to gas
Combinations of elements:
divided into smaller parts; two or more elements join together; by chemical means; 2 or more compounds; mixture of 2 or more liquids

Tapescript
Presenter:
Lesson 8 Describing states of matter
C Listen to the mini-lecture about matter and try to complete the notes.

Lecturer: Matter exists in three states – solid, liquid or gas. Solids have a definite volume and shape, as the molecules are held together by strong forces. Liquids have a definite volume, but the forces holding the molecules together are weaker than the forces in solids, and so liquids do not have a definite shape. The forces holding molecules together in gases are even weaker than the forces in liquids, and so gases have no definite volume or shape. Matter can change from one state to another by giving the molecules more energy and making them vibrate more than normal. This energy is usually provided in the form of heat. As the temperature gets hotter, a substance will change from a solid to a liquid, and from a liquid into a gas. All matter is made from elements – substances that cannot be divided into smaller parts. If two or more elements join together, the resulting substance is called a compound. Compounds cannot be divided into individual elements by solely mechanical means. They have to be divided by chemical means. A mixture consists of two or more compounds which are not joined chemically and can be divided by physical means. A solution is a mixture of two or more liquids, or is when a solid is dissolved in a liquid. In this case, the dissolved solid is called a solute, and the liquid is called a solvent.

D Answer questions about matter

- Set the comprehension task for individual work and pairwork checking.
- Conduct feedback.

Answers

1   Three: gas, liquid and solids.
2   Solids have definite volume and shape.
3   When heat energy is applied to it.
4   The two elements in a compound can only be separated by chemical means, whereas the two elements in a mixture can be separated by physical means.
5   A solute is a dissolved solid, whereas a solvent is the liquid that it dissolves in.

E Use information from a text to complete a table

- Revise how to pronounce numbers by asking the whole class to read the numbers in the box.
- Set the task for individual work. Remind students not to show their completed tables to their partners.

F Ask and answer questions to check numbers in a table

- Read through the speech bubbles. Ask students to ask and answer questions about the numbers they have filled in with a partner.
- Feed back answers to the whole group. If necessary, elicit an explanation for numbers in the calcium sulphate column.

Answers

Col 1: 354 **Col 2:** 2,040 **Col 3:** 2.10
Col 4: 366; 2.11 **Col 5:** 370; 2,600
Col 6: 2.03

CLOSURE

- Ask students to tell you what they can now say in English about matter that they couldn't at the beginning of the lesson.

Lesson 9: Talking about oil products

Objective

- to review vocabulary for different oil products

Vocabulary

- words related to oil products

LEAD-IN

- Books closed. Ask students what different products are made from petroleum. Write suggestions on the board.

A Categorize different oil products

- Books open. Refer students to the diagram and the words in the box. Check understanding of the headings such as *lubricants* and *paving and roofing*.

- Set the task for pairwork. Monitor, but do not conduct feedback at this point. If students have problems with this, tell them to look up the different products in the glossary.

B Read a text to check answers

- Set the reading and checking task for individual work. Encourage students to scan for specific information.

- Conduct feedback. Elicit answers from the whole group. Check pronunciation, particularly of the consonant clusters.

- Ask students what the following refer to in the text: *useless, 88%, 12%, green gas, 40%*.

Answers

Lubricants: lubricating oil
Paving and roofing: asphalt
Fuels: kerosene; industrial fuel oil; LPG, diesel; petrol (gasoline)
Other products: plastics; paraffin wax; fertilizers

C Read to complete a table

- Set the task for pairwork.

- Conduct feedback. Ask for answers from different pairs.

Answers

Oil product	Uses
petrol/gasoline	fuel for cars
diesel	fuel for larger vehicles
LPG	fuel for 'green' cars; domestic heating systems; portable stoves
fuel oil	ships; factories; furnaces
engine oil	used as a lubricant
grease	protects sealed bearings
multigrade lubricants	ensure moving parts in a machine work smoothly

Oil product	Properties
petrol/ gasoline	thick; black; hundreds of compounds
diesel	operates at higher pressures than petrol; more efficient and economical
LPG	versatile; fewer emissions than petrol or diesel
fuel oil	heavier than diesel and petrol
engine oil	less viscous; used as a lubricant
grease	thick
multigrade lubricants	operate at different temperatures without being affected

D Discuss proportions shown on a diagram

- Revise expressing fractions and percentages by writing a sample on the board and asking students to read them correctly.
- Revise passive forms by looking at some examples from the text, e.g., *12% is converted into other materials.* Ask students to underline some more examples.
- Refer students to the diagram. Explain what it represents.
- Ask a student to read the speech bubble to show the type of comments required in the discussion work.
- Set the task for pairwork.
- Conduct feedback. Ask for comments and answers from the whole group.

Answers
4 – jet fuel (kerosene)
19.7 – gasoline

CLOSURE

- Elicit what students know about kerosene and/or other fuels. How do students think the use of oil will change in the future?

Lesson 10: Talking about fractional distillation

Objective

- to listen to a description and discuss the fractional distillation process

Vocabulary

- words related to the fractional distillation process

LEAD-IN

- Review the different types of fuels from the previous lesson.
- Write the words *fractional distillation* on the board. Elicit what it means, then ask students to check by looking in the glossary.

A Discuss questions related to the importance of oil

- Set for pairwork. Put students in small groups and read through the questions with them.
- Set the discussion task and circulate to listen, contribute and encourage discussion.
- Conduct feedback. Ask for opinions and ideas from the different groups. Continue the discussion with the whole group.

B Discuss the process of fractional distillation

- Remind students of their definition for fractional distillation at the beginning of the lesson. Refer them to the diagram and the word box.
- Elicit or check the meaning of *bubble caps*, *condense*, *column* and *trays*.
- Set the discussion task for pairwork.
- Conduct feedback. Elicit ideas from the whole group. Encourage students to guess what happens if they do not know.

C ◆ (CD2 T18) Listen to identify the main topic

- Tell students they are going to listen to someone describing the fractional distillation process. The first time they listen, they should simply say which of the questions in exercise A is answered.
- Play the recording for students to listen and identify the relevant question.
- Conduct feedback. Elicit answers from the class.

Answer

3   How is crude oil changed into different products?

Tapescript
Presenter:
Lesson 10 Talking about fractional distillation
C Listen to someone explaining the fractional distillation process. Which of the questions in exercise A does he answer?

Voice: Crude oil is sometimes exported without being treated. However, crude oil contains hundreds of different kinds of hydrocarbons and they are all mixed up together. It needs to be separated into different products or 'fractions' in order to be much use to us. Fortunately, it isn't too difficult to separate the different hydrocarbons out from each other. This process is known as fractional distillation. It involves heating the oil and separating it into different fractions. This is a relatively simple process because each hydrocarbon has a different boiling point. First, the crude oil or petroleum is heated to a high temperature – about 600 degrees Celsius – in a steam boiler or furnace. When it boils, the liquid oil becomes vapour. The vapour enters the bottom of a long column. The column is much hotter at the bottom (400 to 600 degrees) than at the top (about 20 degrees). It has trays inside it with holes or bubble caps that allow vapour to pass through them. The hot vapour rises through the column, and as it does so, it cools and condenses. Different fractions, components of the oil, condense at different temperatures, in other words, they form at different heights in the column. Substances with high boiling points are at the bottom and substances with low boiling points are at the top, so solid residuals such as wax and asphalt condense at the bottom of the column. Heavy industrial fuel oils condense on the trays above them. Gas oil or diesel distillate is collected higher in the column, kerosene above that and gasoline above that. Petroleum gas is collected at the top of the column. After collection, the liquid fractions can be cooled further and then put into storage tanks, or they can be taken to other areas for chemical processing. In fact this is what usually happens. Chemical processing can actually change some fractions by breaking down the hydrocarbon chains. The yield of gasoline, for instance, can be increased, as heavier fuel oils can be further processed to form gasoline.

D 🔊 (CD2 T19) Listen again for detailed information

- Read through the statements with students to check understanding. Ask them to indicate which they think are true or false.

- Play the recording for students to listen and check/answer the questions.

- Students compare answers.

- Conduct feedback. Elicit the reasons why they have indicated that some are false.

Answers

1 T 2 F 3 F 4 T 5 F 6 F 7 F 8 T

Tapescript
Presenter:
D **Listen again and decide whether the statements below are true or false.**
[REPEAT OF EXERCISE C]

E Describe the distillation process

- Set the reordering task for pairwork. Check before students continue with the written description.

- Set the writing task for individual work and pairwork checking. Emphasize that they should use their own ideas, but that the words in the box in exercise B will help them.

- Individual students read their descriptions to the class.

Answers

1 3 2 1 3 4 4 2

Example answers

5   The hot vapour rises through the bubble caps in the column.

6   As it rises and cools, different fractions condense at different temperatures and different heights in the column. The condensed liquid is collected in the trays.

7   The cooled liquid fractions are stored in special tanks. Some are taken for further chemical processing.

CLOSURE

- Books closed. Ask students to describe the diagram in exercise B.

- Write the words *column*, *vapour*, *trays*, *bubble caps*, *condense* and *boil* on the board. Ask students to give you a sentence using each one.

Answers

A

1 Less.
2 Less.
3 This is subjective. It is probably titanium.
4 This depends on the situation. It is normally quite a lot more difficult.
5 Butane.
6 No, it consumes more than 25% of the world's total oil production.
7 More difficult.
8 False. Only petroleum.
9 Methane is lighter.
10 Both can be piped or transported by tanker.

B **C**

Students' own answers.

D

Lubricant	Advantages	Disadvantages
Grease	• Suitable for high temperatures. • Seals out contaminants. • Can be used in difficult-to-access areas. • Easier to retain than oil.	• Difficult to check amount. • Can block small openings. • Can't be replaced easily.
Oil	• Can be cooled. • Can be replaced easily. • Can be recirculated. • Can pass through small openings.	• Often needs filters and coolers. • Can be contaminated/Needs frequent replacing. • Not suitable for very high temperatures.

• Refer students to the self-assessment grids.

Word list

abrasive
acid
alcohol
aluminium
asphalt
blend
boil
brittle
brittleness
bypass
casing
column
combustible
combustion
compound
condense
conductive
conductivity
copper
corrosive
crack
crude oil
debris
diesel
ductile
durability
durable
elastic
elevation
environment
foam
formation
friction
fusibility
fusible
gasoline

iron
jet fuel
kerosene
LPG
lubrication
lustre
lustrous
malleable
mercury
mixture
multigrade lubricant
paraffin wax
pentane
petrol
petroleum
pigging
propel
recovery
rubber
seal
solid
solute
solvent
sophisticated
steel
string
surface
tensile strength
thermal
traditional
tray
vapour
versatile
vibrate
viscous

DESCRIBING PROCESSES AND PROCEDURES

Lesson 1: Sequencing simple processes

Objective

- to review sequencing language for simple processes

Language

- the imperative
- sequencing words

Vocabulary

- personal and family information

LEAD-IN

- Tell students to imagine you cannot drive a car. Ask them to tell you what to do from opening the car door. Mime what they tell you.

A Identify a process and complete instructions

- Ask students to read through the instructions and say what they think they are for.

- Refer them to the word box. Set the task for individual work and pairwork checking. Point out that they can change the form of the words in the box, e.g., adjective – verb.

- Conduct feedback. Ask for completed instructions from different students. Ask which words needed to change form, and why.

- Ask students how often they need to change a fuse.

- Without looking back at the instructions, ask if they can remember how each instruction began. Write *first*, *second*, etc. on the board.

Answers

How to *change a fuse*

1   First, *ensure* the device is power isolated.
2   Second, use a screwdriver to remove the holding screws from the plug and remove the *cover*.
3   Third, check that the screws holding the wires in place are *secure/tight*.
4   Next, *loosen* the screws which hold the fuse in place and remove the fuse.
5   *Replace* it with a new fuse of the same *voltage*.
6   Then, *tighten* the screws which hold it in place.
7   After that, *replace* the cover and *secure* it with the holding screws.
8   *Finally*, check the device now works.

B Write instructions using the imperative

- Exercise covered. Elicit how to change a wheel on a car. As students require vocabulary, supply it and put it on the board.

- Refer students to the visual and text. Set the task for individual work and pairwork checking. Remind them to use the correct sequencers.

- Conduct feedback. Ask some students to read their instructions to the class.
- Ask how often students have to change a wheel.

Example answers

How to change a wheel

1   First, use a wrench to loosen the wheel nuts.
2   Second, use a jack to raise the car.
3   Then, remove the wheel nuts.
4   After that, remove the wheel.
5   Next, put the new wheel in place.
6   After that, replace the wheel nuts.
7   Lower the car.
8   Finally, tighten the wheel nuts.

C Describe order for putting on PPE

- Remind students of different items of PPE.
- Refer students to the word box. Set the ordering task for pairwork.
- Conduct feedback. Remind them to use the sequencers. Do not confirm answers at this point.

D 🔊 (CD2 T20) Listen to check answers

- Play the recording for students to listen and check answers.

Answers

First, put your legs in the overalls.
Second, put your arms in the overalls.
Third, fasten the overalls.
Next, put on your safety boots.
Then, put on your ear protectors.
After that, put on your safety glasses.
Then, put on your hard hat.
Finally, put on your safety gloves.

Tapescript
Presenter:
Unit 8 Describing processes and procedures
Lesson 1 Sequencing simple processes
D Listen and check your answers.

Voice: First, put your legs in the overalls.
   Second, put your arms in the overalls.
   Third, fasten the overalls.
   Next, put on your safety boots.
   Then, put on your ear protectors.
   After that, put on your safety glasses.
   Then, put on your hard hat.
   Finally, put on your safety gloves.

E Give instructions for lifting a heavy load

- Ask if students often have to lift heavy loads in their work. What and why? Ask how to lift safely.
- Refer them to the visuals and prompts. Set the task for pairwork.
- Conduct feedback. Ask students to tell their suggestions to the class.

CLOSURE

- Elicit which verbs from the lesson can be used for giving instructions.
- Ask students to tell you again how to start driving a car. This time it should be much clearer!

Lesson 2: Talking about safety procedures and electricity

Objective

- to extend vocabulary for describing safety procedures

Vocabulary

- words related to first-aid
- sequencers
- verbs related to procedures for energy isolation

LEAD-IN

- Discuss safety training with the class. Ask who has trained, how often people should be trained, what they learn and how useful the training is.
- Review types of injuries.

A Find words related to injuries and accidents

- Refer students to the wordsearch. Set the task for pairwork.
- Elicit answers. Ask students to spell them for you to write on the board.

Answers

1 Thomas had an electric *shock* and suffered minor *burns*.
2 Bob went on a *first-aid* course.
3 Have you had any *safety* training?
4 The *casualty* was *unconscious* and his *pulse* was weak.
5 The *treatment* helped Ahmed *recover*.
6 A first-aider should assess the situation and *diagnose* the problem.

A	R	T	D	S	P	W	C	D	G	H	K
M	F	H	E	O	R	E	T	E	C	A	U
R	I	P	L	A	S	H	O	C	K	R	N
O	R	S	E	A	N	E	R	A	U	E	C
U	S	X	T	I	N	G	U	S	S	C	O
M	T	A	I	D	I	B	T	U	H	O	N
S	A	F	E	T	Y	N	W	A	J	V	S
A	I	X	B	R	M	P	U	L	S	E	C
B	D	S	P	E	L	W	B	T	N	R	I
S	S	H	P	A	E	N	T	Y	M	O	O
U	I	A	N	T	I	D	C	R	H	N	U
T	O	L	O	M	A	I	B	U	R	N	S
A	N	L	N	E	M	O	N	I	T	O	R
D	I	A	G	N	O	S	E	T	I	C	E
B	D	W	T	T	N	Z	T	S	L	K	R
P	Z	M	N	Y	E	X	Q	L	K	N	I
F	W	E	T	F	K	C	O	A	N	B	U
G	N	I	R	Q	D	Y	M	S	O	P	F

B Discuss questions related to safety and electrical equipment

- Put students in groups of three or four. Set the task for group work.
- Conduct feedback. Elicit ideas and answers from the whole group.

C Describe the procedure for energy isolation

- Elicit what students know about the procedure for energy isolation (lockouts and tags).
- Set the matching task for individual work and pairwork checking.
- Conduct feedback, but do not confirm answers at this point.

D 🔊 (CD2 T21) Listen to check answers

- Play the recording for students to check answers. Elicit the name of the procedure.

Answers

1 Inform all parties of the work to be done.
2 Turn off the point of operation of the device.
3 Turn off the main disconnect switch.
4 Lock the main disconnect switch.
5 Apply a warning tag to the main disconnect switch.
6 Test the isolation.
7 Conduct the maintenance work or inspection.

The procedure is called lockouts and tags.

Tapescript
Presenter:
Lesson 2 Talking about safety procedures and electricity
D Listen and check your answers. What is the name of the procedure?

Voice: If a piece of equipment is undergoing maintenance or inspection, it must be power isolated to protect personnel from injury. The process of doing this involves a system of lockouts and tags. This is the standard procedure. First, you must inform all parties of the work to be done. Then, turn off the point of operation of the device. After that, turn off the main disconnect switch and make sure you lock it. When you lock the disconnect switch, you must then apply a warning tag to it. Then, before you conduct the maintenance work or inspection, you must test the isolation.

E Use correct time words to complete instructions

- See which sequencing words students remember from exercise E, Lesson 1.
- Refer students to the word box. Elicit which word is often used with continuous actions and tenses (*while*) and which word is used with a noun phrase (*during*).
- Set the task for individual work and pairwork checking. Point out that they can use words more than once.
- Ask individual students to read a sentence each to check answers.
- Give an example using *while*, e.g., *While I was working in London, I met my future wife.* Elicit endings for some sentences using *while*, e.g., *While I was repairing my car, ... While I was teaching the class, ...*
- Do the same with *during*. Focus on the nouns that can follow, e.g., *my visit, the inspection, the shift*, etc.

Answers

1 before 2 While 3 before 4 after

1 while 2 during 3 when

CLOSURE

- Ask students to give you instructions on how to drive a car again – this time using *before, after, when, while* and *during*.

Lesson 3: Giving first-aid

Objectives

- to listen to and summarize key points in a talk about first-aid
- to extend use of time expressions and use them to talk about safety

Language

- time expressions

Vocabulary

- words and collocations related to first-aid

LEAD-IN

- Ask students who is trained in first-aid and if they have ever used their training. Is this training interesting/useful? What can they/can't they do?

A Make collocations related to first-aid

- Refer students to the visuals and elicit what the people are doing.
- Set the matching task for pairwork.
- Conduct feedback. Elicit answers from the whole group.

Answers

1 preserve life
2 call for help
3 promote recovery
4 diagnose the problem
5 give treatment
6 assess the situation

B Discuss the work of a first-aider

- Ask students to discuss the two questions in pairs. Encourage them to use collocations from exercise A.
- Conduct feedback. Ask for comments from the whole group.

C 🔊 (CD2 T22) Listen for detailed information

- Set the task for individual work and pairwork checking. Remind students that they do not need to write full sentences.
- Play the recording for students to listen first without making detailed notes.
- Students discuss the questions in pairs. Then play the recording again. Pause the recording at relevant times to allow students to make notes.
- Students check in pairs.
- Ask students to use their notes to give more details to the whole class.

Answers

Aims of first-aid: to preserve life; prevent condition of casualty from getting worse; promote recovery

Responsibilities of a first-aider: assess situation; diagnose problem; give immediate treatment; arrange for medical help

Tapescript
Presenter:
Lesson 3 Giving first-aid
C Listen and make notes.

Voice: First-aid is assistance or treatment given to a casualty for an injury or sudden illness. It is the first assistance given before an ambulance or qualified medical expert arrives. First-aiders may need to use whatever facilities and materials are available at the time. First-aid has three aims: first, to preserve life; second, to prevent the condition getting worse; third, to promote recovery. First-aiders have several responsibilities. They have to assess the situation: it is important to check there is no more danger to the casualty and that the first-aider's life is not endangered. They also have to diagnose the problem, or in other words, try to identify the disease or condition that the casualty is suffering from. Then, they have to give immediate appropriate treatment. This may involve bandaging a wound, putting the casualty into the recovery position or giving artificial ventilation. Finally, the first-aider must arrange for qualified medical staff to attend to the casualty. They may have to call or telephone for help, and then, when the medical services arrive, they need to report to the medical staff. So, to recap, if there is an accident, it is important for the first-aider to do the following things. First, check the area is safe. Second, examine the casualty. Third, try to provide medical help. Fourth, call for help and report to the medical staff. Finally, when the medical services take over, the first-aider can leave the scene of the accident.

D ◉ (CD2 T23) Write instructions and listen to check

- Set for pairwork, then individual work. Ask students to discuss the instructions in pairs, then write them individually. Point out that the sequence is important.

- Play the recording again for students to listen and check.

Answers

1st: check area is safe
2nd: examine casualty
3rd: try to provide medical help
4th: call for help
5th: report to qualified medical staff
6th: when medical staff arrive, leave the scene

Tapescript
Presenter:
🄳 Write instructions for what you should do if you see an accident. Listen again and check you have the correct sequence.
[REPEAT OF EXERCISE C]

E Use correct time expressions

- Write an incorrect sentence on the board, e.g., *The yesterday I saw an accident.* Elicit the correction. Go over which time expressions need the definite article, e.g., *the first/last time*, *the next day/day before*. Point out that students often make mistakes with this.

- Go through the rules for use using *for* + length of time, *since* + point of time in the past and *during* + period of time.

- Set the correction task for individual work. Check answers with the whole group.

Answers

2 last week 4 for fifteen minutes 6 Later in the afternoon 8 The day before yesterday

F Use time expressions when talking about experiences

- Read through the time expressions in the word box. Elicit an example for each from different students – these can be general information, e.g., *Several years ago, my family moved to … .*

- Put students in pairs to talk about their experiences of accidents or first-aid. Encourage them to use the time expressions.

- Conduct feedback. Ask pairs to share their experiences with the rest of the class.

CLOSURE

- Books closed. Revise the collocations from exercise A by giving one word to prompt the second.

- Ask students to list the time expressions.

Lesson 4: Using the ABC rule

Objective

• to read detailed first-aid instructions and sequence them

Vocabulary

• words related to resuscitation

LEAD-IN

• Write the word *resuscitation* on the board. Elicit meaning.

• Ask how people who are not breathing can be resuscitated.

A Review vocabulary for body parts

• Elicit the names for body parts related to the head, neck and torso.

• Books open. Refer students to the table and word box.

• Set the task for individual work and pairwork checking.

• Conduct feedback. Elicit answers from the whole group.

Answers

Parts of the head and neck area:
throat
forehead
chin
lips
Parts of the torso:
ribs
chest
lungs
heart

B Study verbs used in describing resuscitation procedures

• Refer students to the visuals and word box. Set the task for pairwork.

• Conduct feedback. Check understanding of the verbs in the box.

• Make sure students understand the different verbs by asking them for examples of other occasions when they might *kneel*, *tilt*, *blow*, *pinch* and *check*. Ask if there are other verbs they could use in describing the pictures.

• Elicit what procedure students think is happening in each picture.

Answers

1 kneel; tilt
2 pinch; blow
3 check
4 compress

C Read to check answers

• Ask students to read the ABC rule to see whether they were right about the procedures.

• Conduct feedback.

• Check the meanings of *breathe*, *beat*, *circulate*, *respire*, *block*, *clear* and *ventilate*. Ask what the nouns are from some of these verbs.

• Ask students to find two words in the text that mean *very important* (*vital* and *essential*).

D Put procedures in the correct order

- Read through the procedures and check any outstanding unfamiliar vocabulary.

- Set the task for pairwork. Encourage students to look carefully at the time expressions in the text so that they get the correct order.

- Conduct feedback. Elicit answers from the whole group. If students disagree, ask them to give their reasons.

- To review the safety procedures, ask students to close their books and give the instructions from memory.

Answers

1	3
2	9
3	12
4	11
5	7
6	4
7	2
8	10
9	5
10	6
11	1
12	8

E Rewrite sentences with mistakes in

- This exercise is testing students' knowledge of the present perfect. It can be used as a diagnostic exercise. The structure of the present perfect is covered in more depth in the next lesson.

- Point out the use of the present perfect in the first sentence of the final paragraph of the ABC rule text: *After you have given artificial ventilation, ...* Point out the use of *have* as an auxiliary and the past participle. Explain that it is used for past actions where no time is mentioned.

- Tell students that the sentences in exercise E that mention a specific time or date should use the past simple tense. The others should use the present perfect.

- Set the task for pairwork.

- Conduct feedback with the whole class.

Answers

1 I worked offshore in 2001.
2 Have you ever seen an explosion?
3 I started my shift at 7.00 a.m.
4 I haven't worked offshore.
5 Have you been abroad?
6 I have never been in an accident.

CLOSURE

- Books closed. Ask students to tell you the different verbs used in describing artificial ventilation.

- Discuss whether they have ever given this procedure and whether they would be confident should they have to.

Lesson 5: Describing past experiences

Objective

• to review the use of the present perfect to talk about past experiences

Language

• present perfect

LEAD-IN

• Discuss who in the class has done a safety or other practical training course.

A 🔊 (CD2 T24) **Listen to an interview to decide if statements are true or false**

• Read through the statements with students. Set the task for individual work and pairwork checking. Ask students to correct false statements.

• Conduct feedback.

Answers

1 F (It was 3 years ago.)
2 T
3 T
4 F (It was six months ago.)

Tapescript
Presenter:

Lesson 5 Describing past experiences
A Listen to Bob being interviewed about his experience with first-aid training. Decide whether the statements below are true or false.

Interviewer: Do you mind if I ask you about safety training, Bob?
Bob: Sure. What do you want to know?
Interviewer: Well, have you done any safety training before?
Bob: Yes, I have.
Interviewer: When did you do it?

Bob: Ooh, let's see. About three years ago now.
Interviewer: What did you study?
Bob: We studied fire protection and fire injury-related first-aid.
Interviewer: Have you found it useful?
Bob: Yes, … yes, I have.
Interviewer: Have you used the training since then?
Bob: Yes, well, I've used the fire safety, but not the first-aid yet.
Interviewer: Oh, yes? What happened?
Bob: Well, there was a small fire in the warehouse about six months ago and I helped put it out.

B Establish the concept of the present perfect

• Find a student who has done a first-aid course and ask when they did it. Write *A has done a first-aid course. A did a first-aid course … years/months ago* on the board. Elicit which sentence mentions a definite time and which doesn't. See if students remember the name and form of the present perfect from the previous lesson.

• Ask students to look at the sentences in the course book and answer the questions. Ask concept questions, e.g., *Do we know when Bob did the safety training? (Yes.) Do we know when he used the fire safety training? (No.)*

Answers

1 Sentence 1
2 Sentence 2

C Study rules for the past simple and present perfect

- Refer students to the timelines. Set the matching task for pairwork.
- Conduct feedback. Elicit more examples. Ask what cars the students have driven. Ask when they learned to drive, etc.

Answers

The first rule refers to the first timeline. The second rule refers to the second timeline.

D Study the form of the present perfect

- Take one example of the cars students have driven, e.g., *Ahmed has driven a Ferrari. Mehdi, have you driven a Ferrari? (No, I haven't.) Mehdi hasn't driven a Ferrari.* Repeat this. Then set the completion task for pairwork.
- Conduct feedback. Check answers.
- If necessary, drill the form with another set of examples.

Answers

I *have* done first-aid training.
He *has* not *done* first-aid training.
Have you *done* first-aid training?

E ◀) (CD2 T25) Study past participles

- Write a sentence in the past simple on the board, e.g., *I learned to drive a car when I was 20.* Elicit that the verb form is the past simple – one is regular and one irregular. Elicit the past participles. Explain that students learn three parts of a verb: infinitive, past simple and past participle, e.g., *learn–learned–learned, drive–drove–driven.*
- Set the table completion task for individual work and pairwork checking.
- Conduct feedback. Ask students which verbs are regular and which irregular.

Answers

Infinitive	Past simple	Past participle
use	*used*	*used*
work	*worked*	*worked*
study	*studied*	*studied*
do	*did*	*done*
go	*went*	*gone*
fly	*flew*	*flown*
see	*saw*	*seen*

Tapescript
Presenter:
E Write the past simple and past participle of the verbs in the table below. Then listen and check your answers.

Voices:
visit	visited	visited
use	used	used
work	worked	worked
study	studied	studied
do	did	done
go	went	gone
fly	flew	flown
see	saw	seen

F Use past participles to complete sentences

- Set the task for individual work.
- Conduct feeback. Check answers with the whole group.

Answers

1 Ahmed *has done* a first-aid course.
2 Frank *has worked* offshore and he *has flown* in a helicopter many times.
3 Tom *has used* a blowtorch before.
4 *Have* you ever *studied* electronics?
5 I *have been* to Qatar, but I *haven't been to* Kuwait.
6 I *have* never *had* an accident.

CLOSURE

- Books closed. Test students on parts of the verbs from the table in exercise E.

Lesson 6: Talking about events that have/ have not happened

Objective

- to practise using the present perfect for experiences and things that have/have not happened recently

Language

- present perfect with *ever/never/yet* and *already*

Vocabulary

- equipment and devices

LEAD-IN

- Tell students something interesting you did last week, e.g., *I went diving.* Ask if they have ever gone diving. Elicit answers. (*Yes, I have./No, I haven't.*) Lead on from this to other sports and elicit short exchanges. Follow up with *when/where* and examples of the past simple.

A Ask and answer questions in a questionnaire

- Read through the questions in the questionnaire, eliciting full *Have you ever* questions. Look at the speech bubbles at the bottom of the page to show the pattern for the questions and answers. Remind students to use correct time expressions, e.g., *one year ago*, *last year*, *for two years*, etc.

- Ask students to write two more questions in the questionnaire. Set the task for interactive work. Remind students to note down answers.

B Report information from notes

- Ask students to report back detailed information about their partners to the whole group using the present perfect and the past simple, e.g., *Yuri has flown in a plane. He went to Russia two years ago.*

- Assess the percentage of the class that has done the different things.

C Study the use of the present perfect with *yet* and *already*

- Show students a shopping list you have. Tell them that you haven't done the shopping yet, but you plan to do it on the way home this evening. Emphasize *yet*.

- Write a *to do* list from this morning on the board. Include *fill up with petrol*, *mark homework* and *do the shopping*. Indicate with ticks which you have already done and which you haven't done yet.

- Read the two speech bubbles with students. Ask them to answer the questions individually. Set for pairwork checking.

- Conduct feedback.

Answers

1. Mohammed.
2. No.
3. No.
4. Yes.
5. No.
6. When something has happened before now.
7. When something hasn't happened, but will happen in the near future.

D 🔊 (CD2 T26) Listen to identify actions already taken and those not taken yet

- Read through the checklist with students.
- Set the listening task for individual work and pairwork checking.
- Conduct feedback. Encourage the use of full sentences.

Answers

2	Choose the best solution.	✓
3	Complete a risk assessment.	✓
4	Inform people about the job.	✓
5	Select the correct equipment.	✓
6	Isolate the power.	✓
7	Conduct the maintenance (happening now).	✓
8	Test the system.	✗
9	Inform people the job is complete.	✗
10	Reconnect the power.	✗

Tapescript
Presenter:
Lesson 6 Talking about events that have or have not happened
D Listen and complete Vasily's maintenance checklist. Discuss with a partner which things he has already done, and which things he hasn't done yet.

Vasily: Okay, this is the current status on the maintenance of the main gas turbine. We've already identified the problem and chosen the best solution. The risk assessment has been carried out and we've already informed people about the

work. We've also chosen the appropriate equipment and we've power isolated the equipment. We're doing the actual maintenance now. When we have finished, we will do a test on the system. We haven't reconnected the power yet, though.

E Write about a recent activity, course or project

- Ask students to write a *to do* list for themselves for work this week. They should tick and cross the relevant items and tell their partner about what they have done already and what they haven't done yet.
- Ask students to write a few sentences about what they have or haven't done.
- Conduct feedback. Ask some students to read their sentences to the class.

CLOSURE

- Write some points covered in this book on the board. Ask students to say which they have already looked at and which they haven't looked at yet.

Lesson 7: Talking about job interviews

Objective

- to talk about past experiences and activities that started in the past and are continuing now

Language

- present perfect/past simple with *since* and *for*

Vocabulary

- words related to job descriptions

LEAD-IN

- Ask students when they last had or conducted a job interview, what it was for and what questions they were asked. Did they enjoy the interviews? Why/Why not?

A Identify documents related to job interviews

- Set the matching task for individual work.
- Conduct feedback. Check answers with the whole group.
- Check understanding by eliciting the information that each type of document usually contains and which ones students are familiar with.

Answers

There is a CV (Curriculum Vitae), a job ad and a certificate.

B Predict interview questions

- Ask students to read the job ad and discuss in pairs what questions they think Ahmed might be asked.
- Conduct feedback. Elicit suggestions from the whole group.

C (CD2 T27) Listen to an interview for questions asked

- Ask students to first read through the questions.
- Play the recording for students to listen and tick the questions they hear.
- Play the recording again for students to note down Ahmed's answers.
- Students compare answers.
- Conduct feedback. Students give their answers to the whole group.

Answers

1   I've been diving for about six years.
3   I went there two and a half years ago.
5   I've done dive courses and safety and first-aid training quite recently.
7   I originally trained in Kuwait, but I've also done courses in the UK.
8   I've studied English properly for two or three years.

Tapescript
Presenter:
Lesson 7 Talking about job interviews
C Listen and tick the questions that you hear. What are Ahmed's answers?

Interviewer: Well, Mr Abdulkader, can you tell us a bit about yourself? Where are you from originally?

Ahmed: I'm from Algeria, but I've lived in Kuwait and worked in other countries a lot.

Interviewer: How long have you worked in the oil industry?

Ahmed: Well, I've been diving for about six years.

Interviewer: Where did you do your training?

Ahmed: I originally trained in Kuwait, but I've also done courses in the UK.

Interviewer: When were you in Azerbaijan?

Ahmed: I went there two and a half years ago and I've worked on the same rig ever since.

Interviewer: Have you done any in-service training courses?

Ahmed: Yes, I've done dive courses and safety and first-aid training quite recently.

Interviewer: Your English is very good. How long have you studied English?

Ahmed: Thank you! It's not that good, but I've studied English properly for two or three years. I also learned it at school, but I didn't work very hard then!

D Study completed and uncompleted actions

- Direct attention to the two sentences. Elicit which timeline matches which sentence. Ask for differences between the two concepts.
- Give more examples, e.g., *I lived in Qatar for three years. I've lived here for six months.* Elicit some examples from the students regarding periods of work/ subjects they studied/have studied, etc.

Answers

The past simple is used when the period started and finished in the past.
The present perfect is used when the activity is still continuing.

E Use *since* and *for*

- Write *How long have you ...?* on the board. Tell students how long you have been a teacher, giving both a *for* and a *since* example. Put these on the board: *for ten years; since 1997.*

- Elicit some more examples from students. Ask them how long they have studied English, been in the oil and gas industry, been married, etc. Write the times on the board.

- Set the task for individual work and pairwork checking.

- Conduct feedback with the whole group.

Answers

1 *for* ages 2 *for* forty minutes
3 *since* I was born 4 *since* last year
5 *for* three years 6 *since* March
7 *for* two months 8 *since* 1993
9 *since* I started work 10 *for* about an hour
11 *since* two months ago
12 *since* the company began

F Complete rules about *since* and *for*

- Elicit answers to complete the two rules.

Answers

for
since

G Role-play a job interview

- Elicit a few ideas for interview questions before students write their own ideas, individually.

- If possible, pair students from the same fields of work. If not, they should know something about each other's work.

- Set the role-play task for pairwork.

- Circulate and monitor to listen and help where necessary.

- After the role-play, ask some students to repeat their dialogues for the class or report back on how the interview went.

CLOSURE

- Books closed. Ask students what questions Ahmed was asked in his interview.

- Give them a list of time expressions to elicit *since* or *for*.

Lesson 8: Talking about heating and thermostats

Objective

• to expand vocabulary connected with heat and heating systems

Vocabulary

• words related to thermostats

LEAD-IN

• Elicit what students know about thermostats and where they can be found.

A Discuss questions about heating and thermostats

• Read through the questions with students. Set the task for pairwork.

• Conduct feedback. Elicit ideas. Agree on answers. Go over any unknown vocabulary and words that students may have forgotten such as *evaporate*, *vapour* and *boiling point*.

Example answers

1 Metals, e.g., copper.
2 Convection is heat transfer in a gas or liquid by the circulation of currents. Conduction is transfer of heat through a material due to the vibration of atoms.
3 Radiation is the transfer of electromagnetic energy through a medium in the form of waves or rays.
4 When a liquid reaches its boiling point, it evaporates, i.e., it converts to a vapour.
5 The boiling point of a substance is affected by the atmospheric pressure and mixture of impurities.
6 A thermometer measures temperature. A thermostat regulates temperature.

B Read a text to answer questions about how thermostats work

• Ask students to look at the diagram of a thermostat and try to describe how it works. Pre-teach vocabulary such as *invar rod*, *brass* and *expansion*.

• Refer them to the text. Set the reading task for individual work and pairwork checking.

• Conduct feedback. Discuss answers with the whole group.

Answers

1 Brass expands when heated, but invar does not.
2 The expansion of the brass tube pulls the invar rod away from the switch to open it and break the circuit.
3 When the liquid cools, the brass contracts and allows the invar to close the switch and make the circuit.
4 The set temperature is adjusted by changing the tension of the spring that closes the switch.

C Identify parts of a domestic heating system

• Ask students what a domestic heating system consists of.

• Refer them to the diagram. Set the labelling task for pairwork. Emphasize that they only need to label the diagram with the parts in the word box.

• Conduct feedback. Discuss answers with the whole group.

Answers

1   the cold water supply tank
5   the hot water storage tank
8   the boiler

D 🔊 (CD2 T28)   **Listen to a description to label a diagram**

- Play the recording for students to listen and complete the labelling of the diagram.

- Students compare answers.

- Conduct feedback. Draw the diagram on the board and elicit labels for the different parts.

Answers

1   the cold water supply tank
2   the expansion pipe
3   pipe C
4   pipe D
5   the hot water storage tank
6   flow pipe A
7   return pipe B
8   the boiler

Tapescript
Presenter:
Lesson 8 Talking about heating and thermostats
D **Listen and finish labelling the system.**

Voice: A domestic hot water supply system consists of a boiler, a hot water storage tank and a cold water supply tank. These are connected to each other by pipes. Hot water leaves the boiler by flow pipe A at the top of the boiler. This pipe enters the top of the hot water tank. When hot water is taken for domestic use, it leaves the hot water tank from another pipe at the top of the hot water tank: pipe D. This water is replaced by cold water from the cold water supply tank. The water runs down pipe C, which goes from the bottom of the cold water tank to the bottom of the hot water tank. Return pipe B replaces the hot water that leaves the boiler with cooler water. It runs from the bottom of the hot water tank to the boiler. There is also an expansion pipe. This runs from the top of the hot tank and is bent twice at right angles so that its open end is above the cold tank. The expansion tank allows excess hot water steam to run into the cold water tank and lets any air that is in the system escape.

CLOSURE

- Ask students to describe in their own words how a thermostat works.

Lesson 9: Talking about oil and water separation

Objective

- to describe and complete notes about the oil and water separation process

Vocabulary

- words related to the oil and water separation process

LEAD-IN

- Discuss with students why oil-water separation is necessary, and how it can be achieved.

A Complete an explanation with words given

- Refer students to the word box and the explanation. Tell students that they can check the meaning of the terms in the glossary.
- Set the completion task for individual work and pairwork checking.
- Conduct feedback. Ask students to read the explanation aloud to practise pronunciation.

Answers

1   Crude
2   water
3   oil
4   gas
5   heat; emulsion
6   chemical

B Discuss how a heater treater works

- Ask students what a heater treater does.
- Refer them to the diagram and clarify the direction in which the emulsion moves.
- Set the discussion task for pairwork. Indicate that they can use the questions to guide them.

- Conduct feedback. Discuss ideas with the whole group.

Answers

1   Gas is removed first.
2   No.
3   At the first and second stages.
4   Water is heavier.

C Complete the description of a process

- Set the task for individual work and pairwork checking.
- Conduct feedback.

Answers

1   The emulsion is piped into the separator and *the gas is removed* (*through a pipe at the top of the tank*).
2   After *the gas* has been removed, the *emulsion* is heated and *water is produced* (*which is drained through the bottom of the tank*).
3   Then, the emulsion *flows into another section and clean oil flows out* (*of the top of the tank*).

D Describe how a chemical separator works

- Ask students if they know how a chemical separator works.
- Refer them to the diagram. Ask them to discuss the different parts.
- Check meanings and the pronunciation of *inlet*, *weir* and *demulsifying* agent.

- Set the task for individual work and pairwork checking.
- Conduct feedback, but do not confirm answers at this point.

E 🔊 (CD2 T29) **Listen to a description to check answers**

- Play the recording for students to listen and check their answers. Pause after each question.
- Elicit the correct description. Point out the correct prepositions.

Answers

1 The oil and water mixture flows through the inlet valve and accumulates on the left-hand side of the weir.
2 The water sinks to the bottom of the tank and flows out of the water release pipe.
3 The oil floats on the surface of the water, crosses the weir and flows out of the oil release pipe.
4 The pressure of the system is maintained by venting gas through the gas relief pipe and adjusting the relief valves on the water and oil pipes.

Tapescript
Presenter:
Lesson 9 Talking about oil and water separation
E **Listen and compare the description with yours.**

Voice: 1 The oil and water mixture flows through the inlet valve and accumulates on the left-hand side of the weir. That's when the chemicals are added.
2 The water sinks to the bottom of the tank and flows out of the water release pipe.
3 The oil floats on the surface of the water, crosses the weir and flows out of the oil release pipe.
4 The pressure of the system is maintained by venting gas through the gas relief pipe and adjusting the relief valves on the water and oil pipes.

CLOSURE

- Books closed. Ask students to describe the two separating systems in their own words.

Lesson 10: Describing the final stages of separation

Objectives

• to interpret diagrams using appropriate vocabulary
• to give short talks about what happens in the final stages of separation

Vocabulary

• words related to the separation process

LEAD-IN

• Ask students what happens in the final stages of separation. Elicit the word *centrifuge*.

A Swap information to label a centrifuge

• Books closed. Ask students if they know the parts of a centrifuge and can say how it works.

• Books open. Refer students to the diagram. Divide into Student 1 and

Student 2 pairs. Set the information gap task for pairwork. Make sure students are looking at the relevant pages.

• Give students time to read their section of the text and label as much of the diagram as they can.

• They should then give the information they have to their partner. Emphasize that they should not look at their partner's page, but exchange the information orally.

• Conduct feedback. Draw a rough diagram on the board and elicit names for the different parts indicated.

Answers

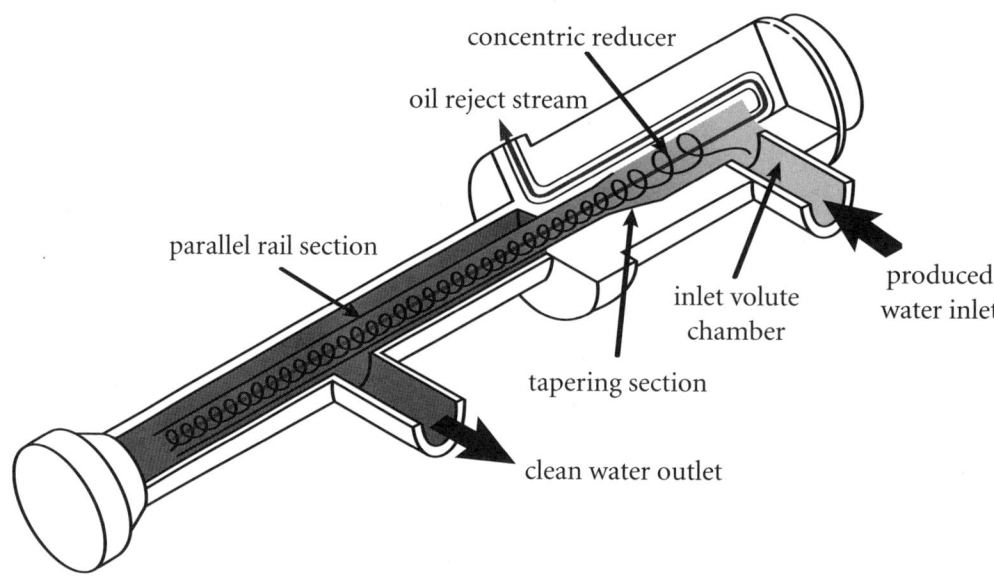

concentric reducer

oil reject stream

parallel rail section

inlet volute chamber

produced water inlet

tapering section

clean water outlet

B Prepare a short talk

- Ask students to read both pieces of text on pages 222 and 232.

- They should now prepare a short talk about how a centrifuge works.

- Point out that they can use the words in the word box as prompts.

- Advise students to make notes for their talk rather than write out the full talk.

- Monitor and give help where necessary.

- Ask some students to give their short talk to the class.

- An extra task could be to ask them to write out the full description.

C Identify tag numbers and abbreviations

- Set the task for pairwork.

- Put pairs together to make small groups. Ask students to compare their ideas.

- Conduct feedback. Share ideas with the whole group.

Answers

1   V1000 = separator
2   PCV = pressure control valve
3   P1001 = pump
4   LCV = level control valve
5   E1000 = compressor suction cooler
6   KO = compressor suction vessel

D Prepare a presentation

- Ask students to work together in their groups to discuss the process and prepare a presentation of how the process works.

- Circulate, encourage and give help where needed.

- Students choose one person in the group to give the presentation to the rest of the class.

CLOSURE

- Books closed. Elicit parts of a centrifuge by writing the first words on the board, i.e., *parallel*, *oil reject*, *concentric*, *clean water*, *inlet volute*, *produced water*.

Review: Describing processes and procedures

Answers

A

Noun	Verb	Adjective
/	*boil*	boiling
compression	*compress*	*compressed*
emulsion	*demulsify*	demulsified
isolation	isolate	*isolated*
separator	*separate*	*separate*
reconnection	*reconnect*	reconnected
recovery	*recover*	*recovered*
conduction	conduct	*conductive*
unconsciousness	/	unconscious
treater/treatment	treat	*treated*
centrifuge	/	centrifugal
breath	*breathe*	/

B

1. compression; recovery; unconsciousness; treatment; breath
2. isolation; reconnection; conduction
3. emulsion; separator; treater; centrifuge

C

1. have been
2. have; been
3. killed
4. is
5. have increased
6. have; started

Assess your skills: Describing processes and procedures

• Refer students to the self-assessment grids.

Word list

airway	invar
artificial ventilation	isolate
assess	jack
atom	junction
blow	kneel
boiler	lip
boiling point	lung
brass	offshore
breathe	operation
breathing	operator
burns	pinch
casualty	preserve
centrifuge	prevent
circulation	promote
compress	pulse
compression	radiation
concentric	reconnect
conduction	recover
contamination	rib
contract	shock
convection	spring
core	thermostat
CV	throat
demulsifier	tilt
diagnose	treatment
disconnect	unconscious
emulsion	vent
examine	voltage
forehead	weir
heart	

GIVING ADVICE

Lesson 1: Talking about transportation

Objectives

- to review vocabulary for transportation
- to introduce a variety of structures that express obligation

Language

- language expressing obligation

Vocabulary

- words related to transportation

LEAD-IN

- Ask students in pairs to name as many types of transportation as they can.
- Conduct feedback with the whole group.

A Identify people and vehicles

- Refer students to the pictures. Ask them to identify the vehicles, machines and people they can see.
- Ask if any students have driven these or any other types of vehicles.

Answers

helicopter
tanker
launch/helicopter pad
forklift truck and driver

B Categorize words related to transportation

- Set the task for pairwork.
- Conduct feedback. Elicit answers from the whole group. List answers on the board, including any other suggestions.
- Discuss with students how important the safety features are and whether, for example, wearing a seat belt is always vital.

Answers

People who are involved with vehicles:
pedestrian; passenger; driver; mechanic
Documentation: registration document; driving licence; insurance certificate
Safety features of vehicles:
ABS; seat belt; airbags; side-impact bars

C Read a text on safety to answer questions

- Before reading the text, ask students if there are any differences between safety regulations for smaller and larger vehicles. Ask what these might be.
- Set the task for individual work and pairwork checking.
- Conduct feedback. Ask students to give their answers to the full group. If they think the statement is true, they should support this with information from the text. If they think it is false, they should correct the statement.
- Ask students to look at the underlined words and guess their meaning. Advise them to use the context – the meaning of the whole sentence – to do this.
- Students compare answers with a partner.

- Conduct feedback. Go through suggestions with the whole group. Confirm guesses.

Answers

1 T
2 F (if possible)
3 F (true for larger vehicles)
4 T (it must be fit for use)
5 F (larger vehicles should have this)
6 F (a driver must ensure the safety of other road users)
7 F (all vehicles should have this)
8 T

D Add words from a text to a table

- Set the task for pairwork.
- Conduct feedback.

Answers

People who are involved with vehicles:
road users
Documentation:
maintenance record
Safety features of vehicles:
air conditioning; first-aid kit;
fire extinguisher; warning triangle

E Study ways of expressing obligation

- Elicit a sentence using *must* by giving an example, e.g., *I need to have a passport to travel abroad. I must have a passport.*

- Read the speech bubbles with students. Set the task for pairwork.

- Conduct feedback. Go through suggestions with the whole group. Confirm correct statements, but do not go into great detail at this stage as there will be more work on modals in the next few units.

- Elicit all the different ways of expressing obligation in the original sentences. Write the structures on the board. Discuss the difference in style and register between the different structures that are used, e.g., *require; necessary;*

mandatory = formal;
need to; have to; must; shouldn't worry about = less formal.

Answers

2 You must have the maintenance record of the vehicle to drive.
3 All vehicles must have side-impact bars.
4 It's necessary for the driver to check if the vehicle is appropriate for the job.
5 You don't have to have air conditioning in larger vehicles.
6 The driver doesn't have to worry about other vehicles on the road.
7 It isn't necessary for smaller vehicles to have a first-aid kit.
8 All vehicles must carry a first-aid kit.

CLOSURE

- Give definitions of new vocabulary from the reading text to elicit correct words, e.g., *You put this on the road if there is an emergency.*

Lesson 2: Talking about rules

Objective

- to practise talking about obligation in the context of transport in the present and past

Language

- present modals: *must/mustn't*; *have to/don't have to*

LEAD-IN

- Set for pairwork. Ask students to write down five rules related to driving.
- Conduct feedback. Discuss the rules with the whole group. How many rules do you have all together?
- Ask students if they think these are good rules or not. Why?

A Discuss statements about vehicles

- Set the task for pairwork.
- Conduct feedback. Check answers with the whole group.

Answers

1 T
2 T
3 T
4 T
5 This depends on the country.
6 T
7 T

B Study the rules for the use of *must* and *have to*

- Ask students to use the statements from exercise A to help them choose the correct modals to complete the rules. They can work in pairs.
- Conduct feedback. Check answers with the whole group.

- Give and elicit extra examples for each rule, e.g., *I must go to bed early tonight* (*I'm tired* – personal obligation). *You must learn the irregular verbs* (my personal obligation on YOU). *I have to teach ten hours a week* (imposed obligation – it's in my contract). *You have to answer five questions in the exam* (imposed obligation – form of the exam).

- Explain that in the positive form, *must* and *have to* are close in meaning, but in the negative they are very different. Contrast *You mustn't smoke in this building* with *You don't have to smoke in this building!*

- Point out that we cannot use *must* for past obligation. We use *had to/didn't have to*.

Answers

Must
Have to
Mustn't
Don't have to
have to
couldn't

C Study the form of *must* and *have to*

- Set the task for individual work and pairwork checking.
- Conduct feedback. Draw the table on the board.

Answers

Present	Necessary	Unnecessary
Affirmative	must *have to*	/
Negative	*mustn't*	don't have to

Past	Necessary	Unnecessary
Affirmative	*had to*	/
Negative	couldn't	*didn't have to*

D Practise the use of *mustn't/don't have to*

- Set the task for individual work and pairwork checking.

- Conduct feedback. Ask why the alternative was not correct. Reinforce the fact that *don't have to* is used when there is no obligation.

- If you think students need extra practice, ask them to write down five things they must do tomorrow, five things they don't have to do tomorrow, then tell their partners.

Answers

1   don't have to
2   mustn't
3   mustn't
4   mustn't
5   don't have to

E Use the correct modal to complete sentences

- Set the task for individual work.

- Conduct feedback. Check answers with the whole group.

Answers

1   has to
2   must
3   have to
4   mustn't
5   don't have to
6   had to/couldn't
7   have to
8   had to

F Invent endings for sentences using modals

- Read the first sentence. Elicit different possible endings from the group.

- Set the task for individual work.

- Conduct feedback. Elicit ideas from the whole group.

- For sentences 1, 3 and 4, ask students to discuss priorities. Which do they think are the most important rules or obligations, and why?

Example answers

1   use their mirrors very carefully
2   carry your driving licence with you
3   block a driver's view
4   using complex controls
5   work very long hours

CLOSURE

- Ask students what they must do after this lesson. Elicit *revise modal verbs*.

Lesson 3: Comparing advice and obligation

Objectives

- to clarify the difference between advice and obligation
- to practise giving advice

Language

- *must*; *have to*; *should*

LEAD-IN

- Ask students what they would say to someone starting a new job. Write their suggestions on the board. Indicate which items are giving advice.

A 🔊 CD2 T30 Listen to a conversation to answer questions

- Set the task for individual work and pairwork checking.
- Give time for students to read through the questions before they listen.
- Play the recording for students to listen and answer the questions.
- Check answers by playing the recording again. Pause after each answer.

Answers

1 He will have to clean up.
2 He has to wear the correct PPE – a hard hat outside and safety boots all the time.
3 He should ask one of the crew and read the regulation handbook.
4 He should read the safety regulation handbook today.
5 He mustn't smoke on the floor of the rig.

Tapescript
Presenter:
Unit 9 Giving advice
Lesson 3 Comparing advice and obligation

A A supervisor is talking to Jim, a new roustabout who is starting work on the floor of the rig. Listen to the conversation and answer the questions.

Supervisor: Okay Jim, at the beginning it won't be very exciting. You'll have to spend a lot of time cleaning up – it's important that the surfaces are not wet or slippery. Make sure you wear the correct PPE.

Jim: Do I have to wear a hard hat and gloves all the time?

Supervisor: Well, you should wear the hard hat in this area mainly, not inside. You don't have to wear gloves all the time – but you must wear the safety boots.

Jim: It's quite a lot to remember!

Supervisor: If you're not sure, you can always ask one of the crew. It's also important to read the regulation handbook.

Jim: I had a look at it yesterday, but I can't remember everything.

Supervisor: Well, try to find time to read it again before you start work. In fact, you should read the safety regulations handbook today if possible.

Jim: I think I need a cigarette.

Supervisor: Well, don't forget, you mustn't smoke on the floor of the rig!

B Understand the differences between *must* and *should*

- Set for pairwork. Ask students to read the sentences and discuss the differences with their partner.
- Conduct feedback. Elicit ideas.

- For extra practice and to help pronunciation of *should/shouldn't*, give some situations and elicit advice from the students, e.g., *I'm very tired. My Arabic isn't very good.* Ask students to repeat the sentences with *should.* Focus on the weak forms.

Answers

Should is used when we give advice and say something is a good idea, but it is not an obligation. *Shouldn't* is used to show that we don't think something is a good idea. Contrast this with *must*, which implies an obligation and necessity, and *mustn't*, which shows something is not allowed.

C Correct mistakes in sentences

- Set the task for individual work and pairwork checking.
- Conduct feedback.

Answers

1 … you should go to bed earlier.
2 You have to obey the safety regulations.
3 … I should do this welding now?
4 … you should ask for time off …
5 We had to wear a life jacket …

D Complete sentences with different modal verbs

- Set the task for individual work and pairwork checking.
- Conduct feedback.

Answers

1 shouldn't 2 mustn't 3 has to
4 should/can 5 have to

E Practise giving advice

- Refer students to the visuals and comments.
- Set for pairwork. Students write an answer for each problem.
- Conduct feedback. Check the answers and advice with the whole group. Discuss whose advice is the best.

F 🔊 CD2 T31 Listen to compare ideas

- Play the recording for students to listen and compare answers.
- For more practice, students can work in pairs to write down some more situations where people need advice. They can then swap situations with another pair and write advice for the problems.

Tapescript
Presenter:
F Listen and check your answers.

Voice 1: He must have been stealing the tools. I'd report him to the supervisor. Yes, I think you should do that.

Voice 2: If you're tired, you should ask if you can take more breaks – just short ones. You should make sure you eat regularly and go to bed early.

Voice 3: The rules are there for a reason. You have to follow them. If you think some of them are unnecessary, perhaps you could ask your supervisor why they are important.

Voice 4: Maybe you should change your job! Or if that's not possible, be patient and see if it gets more interesting. If you show you are good at the job you have, you might be given more interesting things to do. It's a good idea to look around at other people who have the jobs you would like to do and find out what you need to do to get a job like that.

CLOSURE

- Ask students what advice they would give to:
  1 people who want to work in the oil and gas industry.
  2 people who want to do their own particular jobs.

Lesson 4: Troubleshooting

Objective

- to speculate about causes and possible results of problems

Language

- *may/might/could*; *going to/will*

Vocabulary

- words related to problems

LEAD-IN

- Tell students about a problem you have with your car, e.g., *It keeps stalling*. Elicit possible causes.

- Discuss other problems people can have with cars, and possible reasons. Write a couple of sentences on the board with *may/might/could*.

A ◆ (CD2 T32) Make predictions

- Refer students to the visuals and ask what the problems are, e.g., *the plug isn't wired properly*; *there's a lighted cigarette next to an oil drum*; *oil is coming from a truck*.

- Set for pairwork. Students predict what is going to happen.

- Conduct feedback.

- Play the recording for students to listen and compare ideas.

Answers

1   Someone will be electrocuted./It will give someone an electric shock.
2   It's going to catch fire./The cigarette may go out./If there's a strong wind or if the oil is leaking, it might cause a serious problem.
3   It could have an oil leak./It will destroy the engine.

Tapescript
Presenter:
Lesson 4 Troubleshooting
A **Work in groups. Look at the pictures and predict what is going to happen. Listen to some people discussing each picture and see if their ideas are the same as yours.**

1
Voice 1: That plug isn't wired properly, someone will be electrocuted if it isn't fixed.
Voice 2: It's not that dangerous, I think it will give someone an electric shock, but it won't electrocute them!
2
Voice 1: Oh no, the lighted cigarette shouldn't be next to the oil drum – it's going to catch fire!
Voice 2: Yes – it may go out, but if there's a strong wind or if the oil is leaking, it might cause a serious problem.
3
Voice 1: Is that oil on the ground? It may have come from the engine.
Voice 2: Yes, it could have an oil leak. If someone doesn't fix it and top up the oil, it will destroy the engine.

B ◆ (CD2 T33) Discuss likelihood

- Refer students to the box. Play the recording again for them to complete the sentences.

- Conduct feedback. Point out the

differences in the likelihood of the sentences, e.g., 1 is certain (100%), 2 is very likely (90%), 3, 4 and 5 are possible (50% or less). Clarify that there is little difference between 3, 4 and 5, although *could* is often used when things are slightly less likely than *may* or *might*.

- Write some car problems on the board. Elicit how likely it is that the car will break down. Students should use the range of structures.

Answers

1 It's going to *catch fire.*
2 I think it will *give someone an electric shock.*
3 It may *go out.*
4 It might *cause a serious problem.*
5 It could *have an oil leak.*

Tapescript
Presenter:
B Listen again and complete the sentences in the box below.
[REPEAT OF EXERCISE A]

C Look at possible causes for and/ or consequences of situations

- Look at the speech bubbles, then set the task for pairwork.
- Conduct feedback. Elicit ideas from the whole group. Ask for possible results too.

Example answers

1 **Cause:** The outlet could be blocked.
   Consequence: It could cause a flood.
2 **Cause:** The fuse may have blown. The bulb might have gone.
   Consequence: It could be too dark to work.
3 **Cause:** The water supply might be switched off.
   Consequence: There could be a blockage.
4 **Cause:** He might have a cold. He may have forgotten his ear protectors.
   Consequence: It may stop him concentrating.

D Compare modals of possibility

- Set for individual work, then pairwork. Ask students to look at the two sentences, then discuss the questions with a partner.
- Conduct feedback. Explain when we use *may/might have done.* Give some extra practice on this, if necessary, e.g., *Where is Mohammed today? He might have gone on holiday. He may have been delayed in the traffic.*

Answers

1 The second sentence.
2 The first sentence.
3 The second sentence.

E Anticipate problems

- Elicit the word *anticipate.* Check all students understand the meaning.
- Refer students to the visuals. Set the task for pairwork.
- Conduct feedback. Go through explanations and predictions with the class.
- Read through the comment in the box. Ask students for examples of problems they have had recently at work and discuss what was relevant and irrelevant when assessing the problems.

Answers

1 There is damage to the handle of the hammer. The head might fly off and injure someone.
2 The temperature is high. The radiator might have a leak. The car may overheat and break down.
3 A shoelace is undone. The person could trip over the shoelace, fall and injure themselves.

CLOSURE

- Ask students for predictions of how much of this lesson they will remember next time!

Lesson 5: Talking about hazards

Objectives

- to read about and match hazards with advice
- to talk about hazards using active and passive modal verbs

Language

- active and passive forms of modal verbs

Vocabulary

- words related to hazards and rules

LEAD-IN

- Ask students if they, or any of their colleagues, have ever tripped over in the workplace. Ask them to describe what happened.

A Identify slip hazards

- Refer students to the visuals. Elicit what the slip hazards are.

Answers

1   People can trip over the edges of a mat, or the mat itself can slip causing people to fall.
2   A slope can cause people to stumble if they are not expecting it to be there. If it is slippery, it is even more dangerous.
3   Smoke can stop people seeing things on the floor or other items that can cause them to trip or fall.

B Discuss how to minimize hazards

- Set the discussion task for pairwork.
- Conduct feedback. Discuss ideas with the whole group.
- Ask students what else can cause trips and falls in the workplace. Write suggestions on the board.
- Elicit how these can be prevented.

C Match rules and advice

- Check the meanings of *spillage*, *trailing*, *tread indicators*, *hand rails* and *barriers*. Set the task for individual work and pairwork checking.
- Conduct feedback.

Answers

2   a
3   g
4   b
5   c
6   f
7   e
8   d

D Identify active and passive modals

- Remind students of the passive form, e.g., *Peter Banks designed this building. This building was designed by Peter Banks*. Remind them also that we use the correct form of *be* and the past participle.
- Give an example of an active and passive modal, e.g., *Someone must clean this room. This room must be cleaned.*
- Ask students to underline the active modals and circle the passive modals in exercise C. They then compare answers.

- Conduct feedback. Check answers with the whole group. Write some examples on the board.
- Elicit active forms of the passive modals (use *we* as the subject), and vice versa. Ask for full sentences, e.g., *We must position equipment to avoid cables crossing pedestrian routes.*

Answers

Active:
you should ensure; do not have; you should try; you should improve; it is a good idea; you should use

Passive:
equipment must be positioned; this should be eliminated; employees must also be warned; this should be avoided; if it cannot be avoided; it is a good idea; areas have to be kept; rubbish should be removed; these have to be cleaned up; is used

E Make passive sentences

- Set the task for individual work and pairwork checking.
- Conduct feedback. Ask for students to make active sentences from these.
- Elicit when the passive is more appropriate (for official and general rules).

Answers

1   Injuries may be caused by faulty equipment.
2   Safety regulations must be followed at all times.
3   The appropriate tool for the job should always be used.
4   An investigation has to be completed after an incident.

F Construct sentences using active and passive modals

- Set the task for individual work and pairwork checking.
- Conduct feedback. Ask for some sentences from different students.

Example answers

PPE shouldn't be forgotten.
PPE mustn't be defective.
First-aid might not be given in serious situations.
Some employees might not be able to do first-aid.
People must be trained in first-aid.

CLOSURE

- Ask students to make some rules for their own workplaces using passive modals.

Lesson 6: Giving toolbox talks

Objectives

- to extract main points from reading and listening texts that give advice
- to prepare and give a toolbox talk

Language

- active and passive modals

LEAD-IN

- Ask students what sort of explanations they have to give other people in their field of work. Talk about equipment, procedures and different jobs.

A Name fire equipment

- Refer students to the visuals. Ask them to match names and pictures. They can do this in an open group.
- Ask if they have ever used any of these, and to say when and where.
- Elicit the differences between the three types of equipment.

Answers

1 picture 3
2 picture 1
3 picture 2

B Read a toolbox talk to identify the main topic

- Write the words *toolbox talk* on the board. Elicit the meaning.
- Refer students to the *Toolbox talks* information box to check.
- Set the task for individual work and pairwork checking.
- Conduct feedback.

Answer

To give information about fire equipment.

C Read the talk again to answer questions

- Ask students to discuss the questions with a partner.
- Conduct feedback. Students discuss answers with the whole group.

Example answers

1 To give information about uses of different fire extinguishers.
2 Extinguishers containing water should not be used on electrical fires or burning liquids such as gasoline.
3 Information about other fire equipment such as fire blankets, etc.

D 🔊 CD2 T34 Listen to identify the topic of a toolbox talk

- Set the task for individual work.
- Play the recording and ask for answers.
- Play the recording again. Ask some more detailed questions to check understanding, if there is time, e.g., *What lifting machines are mentioned? What is the first rule? Why should you always walk around the machine? What potential hazards might there be? How should you test that the machine can take a load? What rules are there about driving? What should you do at the end of the day?*

- Discuss what made the talk easy (or difficult) to follow (the use of sequencers, pauses). Elicit what type of language was used (imperatives, modals or obligation and advice).

Answer

The talk is about safety procedures for using lifting machines.

Tapescript
Presenter:
Lesson 6 Giving toolbox talks
D **Listen. What is the topic of the talk?**

Voice: I want to talk to you about using cranes, forklifts, mobile plant and other lifting machines. The first rule is that you must have the correct CTA or equivalent licence to use the machinery and you must be authorized by the site. Next, always remember to walk around the machine before you use it to check for problems or defects. Carry out proper checks on the machinery – the tyres, the lights and the brakes particularly. You also have to check for potential hazards such as overhead cables and other employees who might be in the way. Make sure you know the safe working load of your machine and the weight of the load you are going to lift. See if the machine can take the load by lifting it slightly, then halting. But you should never leave the cab while the load is suspended, and you mustn't ever stand under it. When you get in the cab, you must wear a seat belt if it's provided – keep to the speed limit and don't allow unauthorized passengers. Never leave the cab unattended. Lastly, at the end of the day, make sure you park on level ground, and always lock the cab, windows and any covers.

E Prepare a toolbox talk

- Let students choose another more relevant topic for their field of work if they wish. If the topics are too specialized for pairwork, students can prepare individually. By choosing a relevant topic, this should make their talk more interesting for the other students.
- Encourage students to list points in note form on the grid provided.

F Give a toolbox talk

- Change pairs and ask students to use their notes to give their toolbox talk to another partner. Ask some students to repeat their talk for the whole class.

CLOSURE

- Ask students what they can remember about the toolbox talk on lifting equipment.

Lesson 7: Using formal and informal technical expressions

Objective

- to understand and use formal and informal equivalents

Language

- informal and formal language

Vocabulary

- parts of an oil well
- slang expressions

LEAD-IN

- Discuss the parts of an oil well and how it works. Review vocabulary such as *hole, borehole, casing, drill pipe, drill string, kelly, mud system* and *blowout*.

A ◑ (CD2 T35) Listen to label a diagram

- Set the task for individual work and pairwork checking.
- Play the recording for students to listen and label the diagram.
- Conduct feedback.
- Students can check meanings and spelling in the glossary if necessary.

Answers

1 derrick
2 turntable and kelly
3 drill string
4 collar
5 bit

Tapescript
Presenter:
Lesson 7 Using formal and informal technical expressions

A John is a reporter visiting the rig where Bob works. Bob is telling him how the machinery works. Listen and finish labelling the diagram below.

John: We're standing by the rig and I'm with Bob, who has offered to talk to us about it. Bob, can you explain the drilling process to me?

Bob: I'll have a go.

John: Now, I know this is the well with the derrick standing over it.

Bob: Yes, that's right, and you can see the rotating turntable at the top of the well with the kelly.

John: What does the kelly do?

Bob: It's a pipe that transfers the rotary movement to the rotating turntable and the drill.

John: And the drill pipe and collars are called the drill string, is that right?

Bob: Yes, that's right. The drill bit is at the end of the drill string – that's the specialized cutter that cuts up the rock.

B Read a dialogue and identify informal expressions

- Introduce the idea of formal and informal expressions by giving a few examples, e.g., *How are you?* (formal) *How's it going?* (informal) Discuss when and with whom we use formal and informal language.

- Ask students if they know any expressions from their field of work that are informal.
- Students underline relevant expressions and check in pairs.
- Conduct feedback. Ask for the expressions they have underlined. Write them on the board, leaving a space for their equivalents.
- Elicit definitions or more formal equivalents of the words on the board.
- If time allows, let students practise reading the dialogue in pairs.

Answers

dull – worn out
trip the pipe – take the pipe out of the hole to have a look at it
fish/junk – things lost in a hole
fish for it – send a tool down the hole to find lost things
a junk basket – the tool used for this
a round trip – pull the drill string out of the well, then run it back in
downtime – unproductive time

C Identify formal and informal technical expressions in a text

- Elicit or explain what a *blowout* is. Alternatively, ask students to find and read out the definition in the glossary.
- Set the task for individual work and pairwork checking.
- Conduct feedback. Focus on when students are likely to come across formal expressions like this, e.g., in a technical manual or academic essay/article.

Answers

Formal technical expressions:
constructed; locate; accumulations; regulated; utilized; strikes; insufficient; restrain; emerge; divert; ignited
Informal technical expressions:
a blowout; gusher; a kick

D Match formal and standard terms

- Set the task for pairwork.
- Conduct feedback.

- Ask students to test each other in pairs.

Answers

1   ignite – set fire to
2   restrain – hold back
3   strike – hit
4   prevent – stop
5   construct – build
6   risk – danger
7   regulate – control
8   locate – find
9   insufficient – not enough
10 utilize – use

E Categorize informal and standard/formal expressions

- Set the task for pairwork.
- Conduct feedback.
- Elicit when the informal expressions would be used, e.g., when talking to colleagues. Point out that informal language is not generally used when writing reports.

Answers

Formal expressions
It was ignited. The rig is in operation. Can you repair it? He handled the situation.

Informal expressions
We set it ablaze. The rig is onstream. Can you fix it? He kept the situation under control.

CLOSURE

- Books closed. Give students the more formal verbs and expressions and ask them to write down the informal equivalents.
- Conduct feedback. Ask students to spell difficult words aloud.
- Say goodbye to the students in as many informal/formal ways as you can.

Lesson 8: Speaking hypothetically

Objectives

- to listen to a talk about oil recovery for specific information
- to practise talking about unreal situations

Language

- first and second conditionals

Vocabulary

- words related to oil recovery

LEAD-IN

- Ask students what problems there can be when oil wells get older.

A Discuss old oil wells

- Refer students to the visual and see if anyone can explain what it shows. Pre-teach *injection* and *displaced*.
- Set questions for pairwork discussion.
- Conduct feedback.

B 🔊 (CD2 T36) Listen to compare answers

- Set the task for individual work.
- Play the recording for students to listen and complete the task.
- Compare answers.
- Ask students to summarize how water injection works.

Answers

1   The pressure decreases and it gets more difficult to pump.
2   Water.

Tapescript
Presenter:
Lesson 8 Speaking hypothetically
B Listen to Bob explaining about older oil wells. Compare his answers with yours.

Bob: When oil wells get older, it becomes more difficult to pump oil efficiently, as the pressure decreases. The recovery of oil can be improved by using water injection – it can be improved from fifteen percent to as much as fifty percent. Water injection wells are usually located at the periphery or edge of the field and by forcing water below the accumulation, the crude oil is pushed towards the producing wells near the centre. As well as raising the pressure in the wells, water injection also keeps oil below its bubble point. This is the point when gas can escape from the crude oil in great quantities and is produced more than oil. Before water is injected into a well though, it must be treated to remove solids, bacteria and oxygen that might cause damage or corrode the equipment. Sea water can be used for injection on offshore platforms, but it needs to be desalinated – to have the salt removed – because untreated sea water can be highly corrosive.

C 🔊 (CD2 T37) **Listen to decide if statements are true or false**

- Read through the statements with students and elicit any initial comments.
- Play the recording for students to listen and decide which, according to the recording, are true or false.
- Students compare answers with a partner.
- Conduct feedback. Students give you their answers, correcting the false statements.

Answers

1 F (by 50%)
2 T
3 F (if pressure is low, it's difficult to pump efficiently)
4 F (they would produce more gas)
5 T
6 T

Tapescript
Presenter:
C **Listen again and decide whether the statements below are true or false.**
[REPEAT OF EXERCISE B]

D **Identify conditional structures**

- Set the task for pairwork.
- Conduct feedback with the whole group. Get students to read out the sentences, focusing on the pronunciation of the weak forms.
- Give and elicit other examples of unreal/hypothetical situations, e.g., *If I were a lawyer, I would earn a lot more money than I do.* (You aren't, but IF ... .)
- Contrast these with *If I go back to England in the summer, I'll visit my grandparents.* (You might go back, it's not completely sure, but IF ... .)

Answers

First conditional: 1, 2 and 6
The second conditional.

E **Study grammar rules for first and second conditionals**

- Set the task for individual work and pairwork checking.
- Conduct feedback. Go through the completed rules. Check the concept.

Answers

1 inject; improve
2 possible; present; modal
3 used; be
4 unreal; past; would

F **Complete sentences with correct verb forms**

- Set the task for individual work and pairwork checking.
- Conduct feedback.

Answers

1 will
2 would
3 will
4 won't
5 would
6 wouldn't

G **Decide which situations are possible and which unlikely**

- Ask students to discuss the sentences in pairs and decide which are possible and which unlikely for them personally. They should give reasons for their choice.
- Ask them to make conditional sentences from these situations. Do one first with them, e.g., *If my English improves next year, I will earn more money!*
- Conduct feedback. Ask for examples from the whole group.

CLOSURE

- Ask students to finish some sentences that you start, using the appropriate forms, e.g., *If I sell my car ... . If I were on holiday now ... .*

Lesson 9: Talking about drilling mud circulation systems

Objectives
- to read about mud circulation systems
- to talk about the systems speculatively, using conditional structures

Language
- first and second conditionals

Vocabulary
- words related to mud circulation systems

LEAD-IN

- Elicit what students know about drilling mud circulation systems. Are they involved directly with such systems?

A Find phrases and terms used to describe the systems

- Set the task for individual work and pairwork checking. Refer students to the glossary if they are unsure of the answers.

- Conduct feedback. Ask for correct terms. Ask students to describe how these are involved in the system.

Answers

1 rock cuttings
2 drill bit
3 rotary table
4 shale shaker
5 mudman
6 sand trap

B Read and complete a text

- Set the task for individual work and pairwork checking.

- Conduct feedback. Ask one student to read each paragraph aloud and pause at the gaps to allow the class to supply the answers.

- Ask more detailed questions if there is time, e.g., *What does the drilling mud do at the bottom of the well? Why does it bring rock cuttings to the surface? What does the mud consist of? What does the shale shaker do? What does the sand trap do? Where is new mud mixed?*

Answers

1 drill bit
2 rock cuttings
3 rotary table
4 shale shaker
5 sand trap
6 mudman

C Ask and answer questions about the system

- Ask students what a mud pump does. Compare their answers with the exchange in the speech bubbles.

- Refer them to the visual. Set the task for pairwork.

- Conduct feedback. Ask for volunteer questions and answers from the group.

D Discuss possible and hypothetical situations

- Review the form of the two conditional structures and read through the questions. Ask students to identify which are first (1) and which are second (2, 3, 4, 5) conditionals.

- Put students in small groups to discuss the questions.

- Conduct feedback. Ask for ideas from the whole group.

- In groups again, ask students to think of more hypothetical or possible situations related to this system and to write them down. Swap questions with another group and answer them.

- Conduct feedback. Ask questions to the whole group. Elicit answers.

- If an extra activity is necessary, get students to personalize the questions by asking a partner about their own workplaces and jobs starting questions with, e.g., *What would happen if the fire alarm went off where you work?*

Example answers

1   It will need to be replaced.
2   The oil wouldn't pump so efficiently.
3   It would collapse.
4   It would need to be unblocked.
5   It would have to be retrieved (or fished out) and replaced. There would have to be a round trip.

CLOSURE

- Books closed. Ask students to name the different parts of the drilling mud circulation system.

- Ask what problems can happen with this system.

Lesson 10: Describing filters and strainers

Objectives

- to read and listen for detailed information about filters and strainers
- to review modal structures

Language

- modal verbs

Vocabulary

- words related to filters and strainers

LEAD-IN

- Write the words filter and strainer on the board. Ask students where these can be found – at work and/or outside work. Elicit the difference between them. Students can check in the glossary if they are unsure.

A Find opposites

- Set the task for individual work.
- Conduct feedback.

Answers

1   liquid
2   coarse
3   online
4   new
5   decreased

B Complete a text with the correct modal verbs

- Elicit the different modal verbs that the students have studied. Write them on the board. For quick revision purposes, ask for examples of when we use them.
- Ask students to scan the text to identify the topic.
- Set the task for individual work and pairwork checking.

- Conduct feedback. Read the text aloud and pause for students to supply the correct alternatives. Check why the other alternatives are incorrect.
- If time allows, ask students to find words in the text that mean the same as *too much* (*excessive*), *when something doesn't work* (*malfunction*) and *rough* (*coarse*).

Answers

One problem for pipelines is solids in liquids. These solid particles *might* be carried in the liquid or *could* be formed within the pipeline and be the result of corrosion of the pipe wall. As solid particles build up, they lower the flow rate of the pipe. Solid particles *can* also cause excessive wear or equipment malfunction, so they *have to* be removed to protect equipment and extend the working life of parts.

Because of this, filters and strainers *must* be fitted within pipelines. Filters are finer than strainers, so they *may* be used where smaller particles need to be caught. Strainers usually form a metal screen and *can* catch larger coarse particles.

C Read a text to answer detailed questions

- Set the task for individual work.
- Conduct feedback. Check answers with the whole group.

Answers

1 They might be carried in the liquid. They might be the result of corrosion of the pipe wall.
2 They lower the flow rate. They can cause excessive wear and equipment malfunction.
3 Filters are finer than strainers. Filters catch smaller particles.

D Describe and compare different filters

- Ask students if they know the names of any different types of filters and what they are used for.
- Set the task for pairwork.
- Conduct feedback. Go through ideas with the whole goup.

E ◉ ◀) (CD2 T38) Listen to check answers and complete a table

- Play the recording for students to listen and check their answers.
- Conduct feedback.
- Play the recording again, if necessary, for students to complete the table.
- Conduct feedback.

Answers

Cone filters:
Advantages: They are effective/simpler.
Disadvantages: They are not easily accessible for cleaning and repairing.

Duplex filters:
Advantages: They contain more than one filter. There is always one that can be used when the other is being cleaned.
Disadvantages: They are more complex than cone filters.

Tapescript
Presenter:
Lesson 10 Describing filters and strainers
E Listen and check your answers. Complete the table with the advantages and disadvantages of each type of filter.

Voice: Cone filters are fitted within pipelines. They are quite effective and are simpler than duplex filters. The main problem with them is that they cannot be accessed easily and, if they need cleaning or repairing, it is necessary to break the pipeline. Duplex filters contain more than one filter. This means that when one filter is offline for cleaning, the flow can be directed to the other filter.

CLOSURE

- Read the text again, making mistakes with the modal verbs. Encourage students to correct you.

Review: Giving advice

A **B** **C**

Students' own answers.

Assess your skills: Giving advice

• Refer students to the self-assessment grids.

Word list

ablaze	mat
ABS	offline
airbag	online
airlock	onstream
annulus	passenger
ATV	pedestrian
blowout	registration document
borehole	restrain
bubble point	roadworthy
carbon dioxide	rotary hose
construct	rotary table
downtime	round trip
driving licence	sand trap
dull	seat belt
duplex	shale shaker
fish	side-impact bars
gusher	slope
handle	spillage
ignite	strainer
inject	strike
insufficient	toolbox (talk)
junk	trip (the pipe)
kelly	troubleshooting
kick	untreated
locate	utilize
maintenance record	worn

TAPESCRIPTS

 CD1 T1

Presenter:
Unit 1 Giving basic information
Lesson 1 Talking about yourself
A Listen. What is Alan saying? Put the sentences in the right order, A to H.

Alan: Hi, my name's Alan. I'm a technical trainer. I'm married with two children. Let me tell you about my family. My wife's name is Anna. My son, Adam, is 13 and my daughter, Sophie, is 10. They're both students. They aren't with me in Azerbaijan.

 CD1 T2

Presenter:
Lesson 2 Introducing people
B Listen and check your answers. Then practise reading the dialogue in groups of three.

Ahmed: Hello, Bob.
Bob: Hi, Ahmed. How are you?
Ahmed: Fine, thanks. And you?
Bob: I'm very well, thanks.
Ahmed: Do you know Yusef?
Bob: No, I don't.
Ahmed: Bob, this is Yusef. Yusef, this is Bob.
Yusef: Pleased to meet you.
Bob: Pleased to meet you, too. Do you work with Ahmed?
Yusef: No, he's my brother-in-law.

 CD1 T3

Presenter:
Lesson 3 Asking questions
B Listen and check your answers.

Voice 1: What's your name?
Bob: My name's Bob.
Voice 1: How old are you?
Bob: I'm 32.
Voice 1: Are you married?
Bob: Yes, I am.
Voice 1: What's your wife's name?

Bob: It's Helen.
Voice 1: Do you have any children?
Bob: Yes, we have two.
Voice 1: What are their names?
Bob: Paul and Emma.
Voice 1: Where do you live?
Bob: I live in Azerbaijan.
Voice 1: What do you do?
Bob: I'm a trainee operator.

 CD1 T4

Presenter:
Lesson 4: Asking for clarification
B Listen and check your answers.

Voice: 1 Could you speak more slowly, please?
2 Sorry, I don't understand.
3 Could you repeat that, please?
4 What does 'extinguish' mean?
5 How do you spell 'extinguish'?
6 Sorry, I don't know.

 CD1 T5

Presenter:
Lesson 6: Identifying equipment
C Listen and check your answers.

Voice: /eɪ/ play – A – H – J – K
/iː/ see – B – C – D – E – G – P – T – V
/e/ men – F – L – M – N – S – X – Z
/aɪ/ my – I – Y
/əʊ/ go – O
/ɑː/ car – R
/uː/ do – Q – U – W

 CD1 T6

Presenter:
Lesson 8: Talking about shapes and sizes
B Listen and check the pronunciation of your answers.

Voice: 1 a triangle   triangular
2 a rectangle   rectangular

3	a square	square
4	a circle	circular
5	a hexagon	hexagonal
6	a cylinder	cylindrical
7	a sphere	spherical
8	a cube	cuboid

 **CD1 T7**

Presenter:
Lesson 10: Giving definitions
A **Match each job with its definition.**
Then listen and check your answers.

Voice:
1 Divers work under water.
2 A derrick monkey works at the top of the derrick.
3 A doodlebugger works on the seismic crew.
4 Drillers operate the drilling machinery.
5 A floorman works on the floor of the derrick.
6 A jughustler uses geophones on the seismic crew.
7 A metallurgist studies rocks to find oil deposits.
8 Motormen control the drilling engine.
9 A mudman maintains the mud systems.
10 A roughneck assists the driller.
11 Roustabouts do routine cleaning and maintenance.

 **CD1 T8**

Presenter:
Unit 1: Review
C **Listen and check your answers.**

Voice: John is a roughneck on an oil rig. He works on the floor of the rig and helps the driller operate the drilling machinery. He wears a hard hat and overalls. The rig is a semi-submersible rig that is situated in deep waters. It is supported by a pontoon.

 **CD1 T9**

Presenter:
Unit 2 Calculating and measuring
Lesson 1 Saying numbers
B **Listen and check your pronunciation of the numbers.**

Voice:
1 the thirty-first of December, two thousand and six
2 twenty twenty
3 one thousand, three hundred and eighty-nine
4 thirty-three point three percent
5 double oh, double four, two, oh, eight, two, five, double oh, four, oh, three
6 ten sixty-six
7 oh, double seven, five, one, seven, six, three, two, nine, eight
8 a million pounds
9 eleven point three five
10 two hundred and thirty-one square metres

 **CD1 T10**

Presenter:
Lesson 2 Talking about dates and times
B **Listen and check your answers.**

Voice 1: twelve o'clock
Voice 2: twelve o'clock
Voice 1: ten past twelve
Voice 2: twelve ten
Voice 1: quarter past twelve
Voice 2: twelve fifteen
Voice 1: twenty-five past twelve
Voice 2: twelve twenty-five
Voice 1: half past twelve
Voice 2: twelve thirty
Voice 1: twenty to one
Voice 2: twelve forty
Voice 1: quarter to one
Voice 2: twelve forty-five
Voice 1: ten to one
Voice 2: twelve fifty

Presenter:
D Listen and write the times you hear.

Voice: Well, I wake up at six thirty every morning because I start work at eight. The bus collects me at seven fifteen and we get to the terminal just before eight, usually around seven fifty-five. We break for lunch at twelve fifteen and start the afternoon shift at one fifteen. We finish at four fifty because the bus collects us at five. It takes longer to get into the city because of traffic, so I usually get home around six ten.

Presenter:
Lesson 3 Talking about fractions and percentages
E Listen and finish labelling the pie chart.

Voice: Most injuries in the workplace are back injuries. These account for forty-five percent of all injuries. Thirteen percent of injuries are to the arm, and six percent are to the hand, but even more injuries involve fingers and thumbs – sixteen percent, in fact. Leg or lower limb injuries are less common and account for nine percent of all injuries. Eleven percent of injuries are to other parts of the body – that's eight percent to the torso, and three percent to other areas.

Presenter:
Lesson 5 Talking about units of measurement
E Listen and check your answers.

Voice: 1 Five hundred metres is more than one thousand, five hundred feet.
2 One hundred and thirty minutes is more than two hours.
3 Two point nine kilograms is heavier than four pounds.
4 Three feet four inches is more than seventy-eight point nine five centimetres.
5 Twenty-four kilowatts is more than two hundred and forty watts.
6 Seven hundred miles per hour is faster than one thousand kilometres per hour.
7 Nine hundred and fifty millimetres is longer than nought point one metre.
8 Nineteen bar is greater than nine hundred kiloPascal.

Presenter:
Lesson 7 Measuring dimensions
E Listen and compare the descriptions with yours.

Voice: 1 The box is two centimetres high. Its length is six point seven five centimetres and its width is four point two five centimetres.
2 This car is one point four metres high. It is one point eight metres wide and five metres long.
3 The rig is sixty-five point five metres high and its area is thirty-five point five square metres.

Presenter:
Lesson 9 Taking other measurements
C Listen and check your answers.

Voice 1: Do you have all the answers?
Voice 2: I think so, yes.
Voice 1: Okay, so how loud is a noisy factory?
Voice 2: A fairly noisy factory would be one hundred decibels.
Voice 1: And how loud is a chain saw?
Voice 2: A chain saw is one hundred and twenty decibels.
Voice 1: How loud is speech at one metre?
Voice 2: Generally about sixty decibels.
Voice 1: And how loud is a moon rocket at 300 metres?
Voice 2: A moon rocket – that's very loud. It's two hundred decibels.

Voice 1: So how loud is a car horn at 4 metres?

Voice 2: That's eighty decibels.

Voice 1: And the last one. How loud is a quiet office?

Voice 2: Not very loud – that's got to be forty decibels.

 **CD1 T16**

Presenter:

Unit 2 Review

B Listen to someone describing one of the world's largest oil pipelines. Complete the table with the statistics you hear.

Voice: The Baku-Tbilisi-Ceyhan pipeline (sometimes known as the BTC pipeline) transports crude oil one thousand miles from the Azeri-Chirag-Guneshli oil field in the Caspian Sea to the Mediterranean Sea. The length of the pipeline itself is one thousand, seven hundred and sixty kilometres, so it's one of the longest pipelines in the world. It passes through Azerbaijan, Georgia and Turkey. The pipeline has a forty-two-inch diameter for most of its length, narrowing to a thirty-six-inch diameter when it gets near Ceyhan. Normal capacity, from 2009 onwards, is expected to be one million barrels, in other words, one hundred and sixty thousand cubic metres of oil per day. It has a capacity of ten million barrels of oil. It is hoped that the pipeline will be in use for fifty years.

 **CD1 T17**

Presenter:

Unit 3 Describing equipment

Lesson 1 Talking about workshop tools

E Listen and check your answers.

Voice: To make a bench, you need a saw to cut the wood, a vice to hold the wood, a drill to make holes in the wood, and a screwdriver to screw the pieces of wood together.

 **CD1 T18**

Presenter:

Lesson 2 Expressing ability

D Listen and check your answers.

Voice: My brother's an English teacher and he can read and write English really well. I can speak English, but I can't write it. I'm more practical though. I can use a drill and repair things in the house. My brother can't do that.

 **CD1 T19**

Presenter:

Lesson 4 Describing tools

B Listen and check your answers.

Voice: This is an electric drill. You can use it to make holes in wood, metal or concrete. It has a motor inside it, a trigger, a power cord and plug at the bottom, and a chuck and a bit at the front. The power cord supplies power to the motor. The chuck holds the bit in place. The trigger controls the motor. The plug connects the power cord to the power supply.

 **CD1 T20**

Presenter:

E Listen and check your answers.

Voice: This is a pipe wrench. You can use it to rotate pipes. It has a handle, one adjusting nut, a fixed jaw and a moveable jaw. The adjusting nut is behind the moveable jaw and adjusts the position of the moveable jaw.

This is an off-hand grinder. You can use it to recondition tools like screwdrivers and chisels. It has a column, two grinding wheels, safety screens, rests, a control switch and a quenching tank. The control switch on the front of the column operates the grinding wheels. The rests below the wheels hold the work in place, and the screens protect the user from debris. The quenching tank on the front of the column is used to cool the work.

 **CD1 T21**

Presenter:
Lesson 5 Talking about objects
F **Listen and check your answers.**

Voice: 1 The drill is connected to the energy supply by the plug.
2 A file is used by the engineer to finish the metal.
3 The holes are made with an automatic drill.
4 The flow is monitored by a computer.
5 The change in level is shown by a pointer.
6 Pipes can be held in place with a pipe wrench.
7 Liquid in a tank can be measured with a float.
8 The screw cannot be turned clockwise.

 **CD1 T22**

Presenter:
Lesson 6 Describing measuring devices (1): pressure and temperature
E **Listen and check your answers.**

Voice: This is a filled system thermometer. It consists of a bulb, a capillary tube, a bourdon tube, a pointer and a scale. This type of system is completely filled with a liquid, usually mercury. When the mercury in the bulb expands, it goes through the capillary tube and into the bourdon spiral. The spiral uncurls, and this movement makes the pointer move. The pointer then indicates the temperature on the scale.

 **CD1 T23**

Presenter:
Lesson 7 Describing measuring devices (2): level
C **Listen to the description of the device.**

Voice: One method to measure the level of a liquid in a tank is with a float system. A wire connects a float to a counterweight through a system of pulleys. A pointer on the counterweight indicates the level on a scale. As the level of the liquid decreases, the float gets lower and the counterweight is pulled higher. As the level of the liquid increases, the float gets higher and the counterweight is lowered. The change in level is shown by the pointer.

 **CD1 T24**

Presenter:
Lesson 8 Describing how tools work
B **Listen to the description of how a bench drill works, then finish labelling the diagram with the words in the box.**

Voice: This is a bench drill. It can be used to drill holes in wood or metal. It has a drilling table, an operating lever, a chuck, a guard, a locking handle, a motor and controls. The base of the drill is connected to the bench by four bolts. Above this is the drilling table and the locking handle: this is the small handle behind the drilling table. Above this, at the top of the drill, are the controls. They are at the back of the drill, behind the motor housing. The operating lever is on the side of the motor housing and when it is turned clockwise, the drill is lowered. When it is turned anti-clockwise, the drill is raised. The chuck holds the drill bit in place above the drilling table. There is a guard in front of it to protect the user.

 **CD1 T25**

Presenter:
E **Listen and compare your description.**

Voice: This is a hacksaw. It can be used to cut metal. A hacksaw consists of a blade, an adjustable frame and a handle. There is also a blade-tensioning screw joining the blade to the frame. It is turned to make the blade tighter. Another screw connects the frame to the handle and keeps it rigid. This is called the frame-locking screw.

Presenter:
Lesson 10 Describing how pumps work

D Listen and check your answers.

Voice: Reciprocating pump
In this type of pump, the pumping action is produced by the to-and-fro (reciprocating) movement of a piston or plunger within a cylinder. The liquid is drawn into the cylinder through one or more suction valves, then forced out through one or more discharge valves by direct contact with the piston or plunger.

Single-acting pump
When the plunger moves from right to left, the liquid is drawn into the cylinder through the suction ball check. When the plunger reverses and moves from left to right, the liquid is forced out through the discharge ball check. The discharge ball check is forced open by the pressure of the liquid and, at the same time, the suction ball check is forced closed. The movement of the plunger in the cylinder in one direction is called the stroke of the plunger. The distance the plunger moves in and out of the cylinder is called the length of the stroke. Only one side of the plunger takes part in the pumping action, and water is discharged only during one out of every two strokes. For these reasons, the pump is called 'single-acting'.

Presenter:
Unit 4 Giving instructions and warnings
Lesson 1 Following instructions
D Listen and follow the instructions. What do you make?

Voice: 1 Cut a piece of paper three inches by four inches.
2 Draw a horizontal line across the paper one inch from the top.
3 Draw two vertical lines to divide the bottom of the paper into three equal parts.

4 Carefully tear along the vertical lines up to the horizontal line.
5 Fold the left section of the paper up towards you.
6 Fold the right section of the paper up away from you.
7 Put the paper clip on the bottom of the middle section.
8 Hold the paper by the top above your head and drop it!

Presenter:
Lesson 2 Describing controls
D Listen and check your answers.

Voice: 1 Push the lever up to turn the machine on, and push it down to turn the machine off.
2 Pull the trigger to start the drill, and release the trigger to stop the drill.
3 Turn the dial clockwise to increase the flow rate, and turn it anti-clockwise to decrease the flow rate.
4 Press the button on the top to start the machine, and press the button on the bottom to stop the machine.
5 Press the top button to move up, and press the bottom button to move down.
6 Turn the handle clockwise to open the valve, and turn it anti-clockwise to close the valve.

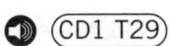

Presenter:
Lesson 3 Giving instructions for using tools
A Bob wants to drill a hole. He asks Vasily for instructions. Listen and put the instructions in the right order, A to M.

Bob: How do I drill a hole in this wall?
Vasily: Well, you need to use a drill. First, measure the work and then mark the hole.
Bob: No problem. What next?

Vasily: Okay, next attach the drill bit and then tighten the chuck.

Bob: Yep, and after that?

Vasily: After that, connect the drill to the power and place the bit over the mark. When you're ready, start the drill and drill the hole.

Bob: What do I do when I finish the hole?

Vasily: First, you stop the drill and secondly, remove the bit from the hole and then disconnect the drill from the power. Finally, you loosen the chuck and remove the drill bit.

Bob: Thanks a lot, that's a great help.

Vasily: My pleasure.

 **CD1 T30**

Presenter:

Lesson 4 Describing and explaining things that are happening now

A Listen to the conversation between Bob and Ahmed. Decide whether the statements below are true or false.

Ahmed: What are you doing, Bob? You don't usually work so late.

Bob: I'm staying late tonight because there's a problem with this forklift truck.

Ahmed: What's the problem?

Bob: The hydraulic mechanism is faulty. This valve isn't working properly.

Ahmed: What are you doing now?

Bob: I'm trying to have a look at it, but it's a bit hot.

Ahmed: Do you need a hand?

Bob: It's okay, there's an engineer coming in the morning.

Ahmed: Right, Bob. See you tomorrow.

Bob: Okay, see you.

 **CD1 T31**

Presenter:

Lesson 5 Giving warnings

B Now listen and tick the sentences you hear.

Bob: Hey, what are you doing? You can't smoke in here! It's dangerous.

New employee: Okay, I'll go outside.

Bob: You're going the wrong way. Don't go in there! It's restricted personnel only.

New employee: Sorry, I'm new here.

Bob: Well, be careful! Always read the safety signs.

 **CD1 T32**

Presenter:

Lesson 6 Comparing temporary and permanent situations

E Listen and check your answers.

Voice: Hassan is studying to be a production engineer. He has a degree, and at the moment he is doing a training course. The course involves studying in the workshop and working on site. He likes the training that happens on site because it shows how things work in action and not just in theory. This week they are studying pigging. He knows some things already, but is finding the course useful, especially when their trainer explains about the different types of pig, and what they do.

 **CD1 T33**

Presenter:

Lesson 7 Talking about problems in the workshop

C Listen. What are the problems?

Vasily: What's the problem?

Alan: It's this machine. It's overheating.

Vasily: You'd better turn it off.

Alan: I can't. The switch isn't working, so the machine isn't stopping.

Vasily: Try the emergency shut down.

Alan: That isn't working either.

Vasily: Now that is a problem! Turn off the power and restart it in ten minutes.

 **CD1 T34**

Presenter:

D Listen again and complete the dialogue. Put the verbs into their correct form.

Vasily: What is the problem?

Alan: The switch doesn't work, so the machine won't stop.

Vasily: Try the emergency shut down.

Alan: That doesn't work either.

Vasily: Turn off the power and restart in ten minutes.

Presenter:

Lesson 8 Talking about the weather

C Listen to the weather forecast and answer the questions.

Voice: Well, we've had some very poor weather conditions in Britain this week due to the cold front that's come in from the Atlantic. There was cloud and thick fog in many parts of Scotland yesterday, and temperatures near freezing. Most areas are quite cold today, but there should be some sunny intervals later, although there could be rain and sleet showers – particularly in the North. In fact, at the moment it's snowing in Aberdeen. Rain over the hills will clear later, but we're expecting a heavy frost, and watch out for icy roads – they could be hazardous tomorrow morning. Temperatures down to minus 3 tonight. It's looking better at the weekend, but strong winds will make it feel cold. That's all from me this morning. Drive carefully and have a good weekend.

Presenter:

Unit 5 Describing systems

Lesson 2 Describing heating systems

C Listen and check your answers.

Voice: Open-loop systems

Open-loop systems are manual control systems. The system is controlled by an operator. The system includes a water heater tank, an indicator, burners and a hand gas control valve. Cold water enters the tank through the 'in' pipe at the bottom. The water is heated by burners below the tank. Hot water rises to the top of the tank, and leaves it by the 'out' pipe. An indicator shows the temperature of the water in the 'out' pipe. The temperature can be adjusted by the operator using the hand gas control valve to change the gas supply to the burners.

Closed-loop systems

Closed-loop systems are automatic control systems; a controller takes the place of the operator. The system consists of a controller, a sensing element and a control valve. The sensing element monitors the flow and sends information to the controller. The flow can be adjusted by the controller sending a signal to the control valve, which regulates the flow.

Presenter:

Lesson 3 Describing alarm systems

B Listen to someone describing alarm systems. Decide whether the statements below are true or false.

Voice: There are two types of alarm – audible alarms and visual alarms, although often both are used together. An audible alarm warns the operator there is a problem, and a visual one can give a more specific idea of where the problem is. Basically, an alarm is an on-off control circuit; one that uses a limit-sensing device connected to a warning device. The alarm will go off when the equipment or process is operating outside the pre-set, normal operating range, for example, if there's too much pressure or heat, or too great a quantity of something. The limit-sensing device can be a pressure, temperature, float-operated or flow-actuated switch. When the limit is reached, the switch contacts close, completing an electrical circuit that activates the alarm.

Presenter:

C Listen again and complete the notes.
[REPEAT OF EXERCISE B]

Presenter:

Lesson 4 Describing how electrical systems work

F Listen and check your answers.

Voice: Lamp A goes on if switch A is in the ON position. Lamp B goes on if switch A and switch B are in the ON position. Lamp B doesn't go on if switch B is in the ON position, but switch A is in the OFF position.

Presenter:

Lesson 7 Using process and instrument drawings

D Listen and check your answers.

Voice: 1 level indicator
        2 temperature indicator
        3 pressure alarm
        4 level alarm
        5 temperature recorder
        6 flow recorder
        7 pressure recorder indicator
        8 speed indicator controller
        9 flow indicator controller

Presenter:

Lesson 8 Using tag numbers

B Listen to the description of the diagram.

Voice: The T-4501 is a water storage tank located on the upper deck. Water is pumped to T-4501 from tank T-3501 located on the lower deck. To fill T-4501, a hose is connected from tank fill nozzle N4 on the bottom of T-4501 to manual valve 3501-MV-028 below. This valve must be open when you begin to fill the tank and also the one to the left of it: 3501-MV-005. You have to open the oil skim nozzle valve 4501-MV-006 at the top of the storage tank and continue to fill the tank up until water overflows from the skim nozzle. After that, you close the

valves and disconnect the hose. The level of the water needs to be marked on level gauge 4501-LG-102.

Presenter:

C Listen to a description of another system. Match the sentence halves.

Voice: The PR monitoring systems are used to monitor pressure. The PR427 series are suitable for measuring high pressures. When the pressure reaches a preset high-high level, the alarm PR427-A alerts the operator. PR427-C automatically closes valves PV576A and B. This leads to a reduction in the flow, and lowers the pressure. The PR427-C has a back-up system PR427-C2. PR427-C2 initiates if PR427-C malfunctions. The operator also has emergency override PR0993 if both PR427-C systems fail.

Presenter:

Lesson 10 Measuring flow

A Listen to the definitions and guess which words in the box they are describing.

Voice: 1 It's a container that's used to hold liquid or gas. It's usually made of metal, plastic or fibreglass.
        2 It's an instrument that allows excess liquid to escape.
        3 It's part of a mechanical or electrical system that rotates.

(CD1 T44)

Presenter:

B Listen again and complete the sentences.

[REPEAT OF EXERCISE A]

CD2 T1

Presenter:
Unit 6 Talking about safety
Lesson 1 Talking about parts of the body and injuries
D **Alan was in an accident. Listen and circle the body parts he hurt.**

Alan: I've had one accident at work, quite a nasty one actually. It was on New Year's day – the first of January, early in the morning. I slipped on a wet floor in the workshop and fell. I hit my head on the corner of the work bench as I fell and cut my head – it was quite a deep laceration. Then I landed badly on my hand and broke my right wrist and sprained my right thumb. I also managed to hurt my back, and I got some nasty bruises on my right arm.

 CD2 T2

Presenter:
E **Listen again and complete the accident report form.**
[REPEAT OF EXERCISE D]

CD2 T3

Presenter:
Lesson 2 Talking about PPE and safety equipment
E **Listen and identify the equipment above that people are discussing.**

Voice: 1 It's a sort of container, often red, that hangs on the wall and has water, foam or chemicals in it for putting out fires.
  2 They're orange and white objects made of plastic. You see them on roads or around any area that is dangerous, to stop people going there.
  3 What's the name of that thing you wear on your face? It's something that's made of rubber or plastic, and stops you inhaling poisonous fumes or gases.

  4 It's something that goes around you to stop you drowning if you fall in the water.

CD2 T4

Presenter:
Lesson 5 Talking about risk assessment
B **Bob wants to move the printer from table A to table B. Listen to his risk assessment and complete the sentences.**

Bob: We're going to have to move things around in here a bit. It's too dangerous at the moment. We need to move the printer, but if we put it on the table over there, it'll probably collapse. And that cable, if we don't move it, someone will trip over it. I know, we'll get rid of that table and move this one over there with the laptop and printer. But first we need to sort out the water dispenser. Someone might slip if we don't mop up the spilt water. And make sure you lift the table carefully. Remember, if you move a heavy object, you must bend your legs before you lift it.

CD2 T5

Presenter:
Lesson 6 Talking about past events
A **Listen to Ahmed and Bob talking. What was the problem?**

Ahmed: Had a good day, Bob?
Bob: Not bad, but we had a bit of a panic this morning.
Ahmed: Why? What happened?
Bob: Well, when I got to work at 7, I heard the alarm.
Ahmed: Oh dear, was it serious?
Bob: Well, I checked the control panel of course, and saw the flow rate was really low. So I opened all the valves to increase the flow rate – but it didn't help much.
Ahmed: So what did you do?
Bob: Oh, I called the pump station and they increased the flow rate. Then it was okay.

Presenter:
G **Listen and check your answers.**

Supervisor: What happened yesterday?
Hassan: I turned on the monitoring equipment and checked the system as usual. Then, at 11 o'clock, an alarm sounded.
Supervisor: What was the matter?
Hassan: The flow rate was too high.
Supervisor: Did you initiate an emergency shutdown?
Hassan: No, I didn't initiate an emergency shutdown. I adjusted the flow rate and lowered the pressure in the pipeline. The alarm stopped when the flow rate reached normal levels.

Presenter:
Lesson 8 Reporting incidents
D **Complete the STOP report cards with the verbs in the correct tense. Then listen and check.**

Voice 1: A technician noticed a piece of wood with nails sticking out in front of the equipment lockup. First, he bent the nails over with a hammer, then put the wood in the garbage container. I congratulated him on his action.
Voice 2: While I was washing my hands, I noticed a broken mirror frame in the bathroom, which was in danger of falling. I informed Administration, and the company repaired the frame.
Voice 3: The light in a classroom was malfunctioning before lunch. The light was flashing on and off. I thought it was a short circuit and a fire hazard. I notified Administration and they immediately sent someone to repair the light.

Presenter:
Lesson 9 Asking about incidents
C **Listen to Bob describing an incident. Think of questions you would like to ask about it.**

Bob: I cut my hand when I was working. I was changing a valve, and while I was removing it, my pipe wrench slipped and cut into my hand. I wrapped my hand in my handkerchief and got someone to take me to the medical block. The doctor said I needed stitches, and I thought 'Oh no!' Then, when he was stitching up my hand, I fainted!

Presenter:
D **Listen to Ahmed asking Bob about the incident. Were his questions the same as yours?**

Ahmed: When did it happen?
Bob: Yesterday evening, just before the end of my shift.
Ahmed: Why were you changing the valve?
Bob: Oh, it was old and faulty.
Ahmed: Who took you to the medical block?
Bob: One of the roustabouts.
Ahmed: Did it hurt?
Bob: Not at first. It started hurting when I had the stitches, and it's quite sore now.
Ahmed: What did the doctor do when you fainted?
Bob: I don't know! When I woke up, I was lying on the couch!

Presenter:
Unit 7 Making comparisons
Lesson 1 Making general comparisons between two things

D **Listen and circle the words and phrases that are used to compare the different types of oil recovery.**

Voice: Nodding donkeys have been around for a long time. They are a traditional way of extracting oil. They are smaller than offshore platforms, and cheaper to run. They are also clearly easier to maintain than offshore platforms. On the other hand, they cannot be used offshore, so offshore platforms were needed when oil started to be extracted out at sea. Offshore platforms are designed to drill more deeply than the traditional nodding donkey. They are more sophisticated and effective, but consequently, they are potentially more dangerous. A nodding donkey has the advantage of usually being more reliable than an offshore platform, but an offshore platform is ultimately more useful.

🔊 (CD2 T11)

Presenter:
Lesson 2 Making more specific comparisons
B **Listen and check your answers.**

Voice: This diagram shows an oil well casing. Casings are used to protect the weaker, upper parts of the oil well, and stop the sides of the well falling in. This is particularly the case with deep formations, as they have higher pressures than shallow formations. When a well is drilled, the top diameters are always larger than those deeper in the well. Consequently, the hole and casing diameter are narrower at the bottom of the well than at the top. The hole is always slightly bigger than the casing, so that a cement bond can be pumped between the outside of the casing and the wall of the hole. After the first section of a well is drilled, a wide diameter casing is fitted inside the hole. A drill bit smaller than the casing is then used to drill the next section of the hole.

🔊 (CD2 T12)

Presenter:
E **Listen and write down two adjectives that are used to describe each liquid, e.g.,** *viscous, heavy*, **etc.**

Voice 1: Crude oil is viscous, much more viscous than water. It's also heavy, isn't it?
Voice 2: It's heavier than water … and oil is flammable. LPG gas is highly combustible, so it's quite dangerous. It's more explosive than oil, for instance.
Voice 1: What about acid? Acid is corrosive. It's also dangerous.

Voice 2: Yes, I think acid is slightly more dangerous than oil or gas.
Voice 1: Water is generally cheap. But it's important if you don't have it – it's essential for life.
Voice 2: Chemically, water is transparent and inert.

🔊 (CD2 T13)

Presenter:
Lesson 3 Comparing more than two things
B **Listen and complete the table above.**

Voice: As regards hardness, rubber is not hard at all. Gold and glass can be classed as hard. Wood is harder than rubber, but steel is the hardest substance. If we compare price, it depends what you're making. In production terms, steel is more expensive than wood, but gold is the most expensive. Rubber is the most elastic substance, although you might be surprised to hear gold is also elastic. Rubber and wood are both combustible substances. Rubber is more combustible than glass, gold or steel, but wood is the most combustible. Wood is probably the most versatile material, but steel is also quite versatile. Glass, rubber and gold are used to make a narrower range of products because they are less versatile. Perhaps rubber is the least versatile.

🔊 (CD2 T14)

Presenter:
Lesson 4 Comparing metals
C **Listen and underline the stressed syllable in each noun in exercise B.**

Voice: 1 durability
          2 ductility
          3 conductivity
          4 fusibility
          5 hardness
          6 brittleness
          7 malleability
          8 lustre

 CD2 T15

Presenter:
G **Listen to the conversation and check your answers.**

Voice 1: Chains need to have tensile strength and hardness. They should be malleable in order to bend the metal into links.

Voice 2: Yes. They also need to resist chemical corrosion and have durability, especially if they are in the open air, on the deck of a rig, or on a boat. They shouldn't break or change their length, so they can't be brittle or ductile.

Voice 1: Storage tanks are durable and can't be too brittle. They have to have hardness and resist surface indentations and corrosion.

Voice 2: What about hot water tanks?

Voice 1: Well, they shouldn't be conductive or fusible. They don't need to have lustre.

Voice 2: The metal for heating elements needs conductivity to heat the air or water around the element. They need to be made of a malleable material to be bent into the correct shape.

Voice 1: Yes, and they also need durability and their surface needs to be hard so that they remain smooth.

 CD2 T16

Presenter:
Lesson 7 Measuring temperature
B **Some thermometers contain mercury, some contain pentane and some contain alcohol. Listen to the comparison of thermometers and complete the table with the names of the three liquids.**

Voice: Thermometers have many industrial, medicinal and domestic uses. A good device is accurate to nought point one degrees Celsius. Typical liquids used in thermometers are mercury, alcohol and pentane. Mercury has the widest range and is particularly good for measuring high temperatures up to five hundred and ten degrees Celsius. Alcohol thermometers have the shortest range, but they can be used to measure temperatures between minus eighty degrees and seventy degrees Celsius. For very low temperatures however, pentane is used, as it can measure temperatures as low as minus two hundred degrees.

 CD2 T17

Presenter:
Lesson 8 Describing states of matter
C **Listen to the mini-lecture about matter and try to complete the notes.**

Lecturer: Matter exists in three states – solid, liquid or gas. Solids have a definite volume and shape, as the molecules are held together by strong forces. Liquids have a definite volume, but the forces holding the molecules together are weaker than the forces in solids, and so liquids do not have a definite shape. The forces holding molecules together in gases are even weaker than the forces in liquids, and so gases have no definite volume or shape. Matter can change from one state to another by giving the molecules more energy and making them vibrate more than normal. This energy is usually provided in the form of heat. As the temperature gets hotter, a substance will change from a solid to a liquid, and from a liquid into a gas. All matter is made from elements – substances that cannot be divided into smaller parts. If two or more elements join together, the resulting substance is called a compound. Compounds cannot be divided into individual elements by solely mechanical means. They have to be divided by chemical means. A mixture consists of two or more compounds which are not joined chemically and can be divided by physical means. A solution is a mixture of two or more liquids, or is when a solid is dissolved in a liquid. In this case, the dissolved solid is called a solute, and the liquid is called a solvent.

 CD2 T18

Presenter:
Lesson 10 Talking about fractional distillation
C **Listen to someone explaining the fractional distillation process. Which of the questions in exercise A does he answer?**

Voice: Crude oil is sometimes exported without being treated. However, crude oil contains hundreds of different kinds of hydrocarbons and they are all mixed up together. It needs to be separated into different products or 'fractions' in order to be much use to us. Fortunately, it isn't too difficult to separate the different hydrocarbons out from each other. This process is known as fractional distillation. It involves heating the oil and separating it into different fractions. This is a relatively simple process because each hydrocarbon has a different boiling point. First, the crude oil or petroleum is heated to a high temperature – about 600 degrees Celsius – in a steam boiler or furnace. When it boils, the liquid oil becomes vapour. The vapour enters the bottom of a long column. The column is much hotter at the bottom (400 to 600 degrees) than at the top (about 20 degrees). It has trays inside it with holes or bubble caps that allow vapour to pass through them. The hot vapour rises through the column, and as it does so, it cools and condenses. Different fractions, components of the oil, condense at different temperatures, in other words, they form at different heights in the column. Substances with high boiling points are at the bottom and substances with low boiling points are at the top, so solid residuals such as wax and asphalt condense at the bottom of the column. Heavy industrial fuel oils condense on the trays above them. Gas oil or diesel distillate is collected higher in the column, kerosene above that and gasoline above that. Petroleum gas is collected at the top of the column. After collection, the liquid fractions can be cooled further and then put into storage tanks, or they can be taken to other areas for chemical processing. In

fact this is what usually happens. Chemical processing can actually change some fractions by breaking down the hydrocarbon chains. The yield of gasoline, for instance, can be increased, as heavier fuel oils can be further processed to form gasoline.

 CD2 T19

Presenter:
D **Listen again and decide whether the statements below are true or false.**
[REPEAT OF EXERCISE C]

 CD2 T20

Presenter:
Unit 8 Describing processes and procedures
Lesson 1 Sequencing simple processes
D **Listen and check your answers.**

Voice: First, put your legs in the overalls.
Second, put your arms in the overalls.
Third, fasten the overalls.
Next, put on your safety boots.
Then, put on your ear protectors.
After that, put on your safety glasses.
Then, put on your hard hat.
Finally, put on your safety gloves.

 CD2 T21

Presenter:
Lesson 2 Talking about safety procedures and electricity
D **Listen and check your answers. What is the name of the procedure?**

Voice: If a piece of equipment is undergoing maintenance or inspection, it must be power isolated to protect personnel from injury. The process of doing this involves a system of lockouts and tags. This is the standard procedure. First, you must inform all parties of the work to be done. Then, turn off the point of operation of the device. After that, turn off the main disconnect switch and make sure you lock it. When you lock the disconnect switch, you must then apply a warning tag to it. Then, before you conduct the maintenance work or inspection, you must test the isolation.

 **CD2 T22**

Presenter:
Lesson 3 Giving first-aid
C **Listen and make notes.**

Voice: First-aid is assistance or treatment given to a casualty for an injury or sudden illness. It is the first assistance given before an ambulance or qualified medical expert arrives. First-aiders may need to use whatever facilities and materials are available at the time. First-aid has three aims: first, to preserve life; second, to prevent the condition getting worse; third, to promote recovery. First-aiders have several responsibilities. They have to assess the situation: it is important to check there is no more danger to the casualty and that the first-aider's life is not endangered. They also have to diagnose the problem, or in other words, try to identify the disease or condition that the casualty is suffering from. Then, they have to give immediate appropriate treatment. This may involve bandaging a wound, putting the casualty into the recovery position or giving artificial ventilation. Finally, the first-aider must arrange for qualified medical staff to attend to the casualty. They may have to call or telephone for help, and then, when the medical services arrive, they need to report to the medical staff. So, to recap, if there is an accident, it is important for the first-aider to do the following things. First, check the area is safe. Second, examine the casualty. Third, try to provide medical help. Fourth, call for help and report to the medical staff. Finally, when the medical services take over, the first-aider can leave the scene of the accident.

 **CD2 T23**

Presenter:
D **Write instructions for what you should do if you see an accident. Listen again and check you have the correct sequence.**
[REPEAT OF EXERCISE C]

 **CD2 T24**

Presenter:
Lesson 5 Describing past experiences
A **Listen to Bob being interviewed about his experience with first-aid training. Decide whether the statements below are true or false.**

Interviewer: Do you mind if I ask you about safety training, Bob?
Bob: Sure. What do you want to know?
Interviewer: Well, have you done any safety training before?
Bob: Yes, I have.
Interviewer: When did you do it?
Bob: Ooh, let's see. About three years ago now.
Interviewer: What did you study?
Bob: We studied fire protection and fire injury-related first-aid.
Interviewer: Have you found it useful?
Bob: Yes, … yes, I have.
Interviewer: Have you used the training since then?
Bob: Yes, well, I've used the fire safety, but not the first-aid yet.
Interviewer: Oh, yes? What happened?
Bob: Well, there was a small fire in the warehouse about six months ago and I helped put it out.

 **CD2 T25**

Presenter:
E **Write the past simple and past participle of the verbs in the table below. Then listen and check your answers.**

Voices:

visit	visited	visited
use	used	used

work	worked	worked
study	studied	studied
do	did	done
go	went	gone
fly	flew	flown
see	saw	seen

 CD2 T26

Presenter:

Lesson 6 Talking about events that have or have not happened

D **Listen and complete Vasily's maintenance checklist. Discuss with a partner which things he has already done, and which things he hasn't done yet.**

Vasily: Okay, this is the current status on the maintenance of the main gas turbine. We've already identified the problem and chosen the best solution. The risk assessment has been carried out and we've already informed people about the work. We've also chosen the appropriate equipment and we've power isolated the equipment. We're doing the actual maintenance now. When we have finished, we will do a test on the system. We haven't reconnected the power yet, though.

 CD2 T27

Presenter:

Lesson 7 Talking about job interviews

C **Listen and tick the questions that you hear. What are Ahmed's answers?**

Interviewer: Well, Mr Abdulkader, can you tell us a bit about yourself? Where are you from originally?

Ahmed: I'm from Algeria, but I've lived in Kuwait and worked in other countries a lot.

Interviewer: How long have you worked in the oil industry?

Ahmed: Well, I've been diving for about six years.

Interviewer: Where did you do your training?

Ahmed: I originally trained in Kuwait, but I've also done courses in the UK.

Interviewer: When were you in Azerbaijan?

Ahmed: I went there two and a half years ago and I've worked on the same rig ever since.

Interviewer: Have you done any in-service training courses?

Ahmed: Yes, I've done dive courses and safety and first-aid training quite recently.

Interviewer: Your English is very good. How long have you studied English?

Ahmed: Thank you! It's not that good, but I've studied English properly for two or three years. I also learned it at school, but I didn't work very hard then!

 CD2 T28

Presenter:

Lesson 8 Talking about heating and thermostats

D **Listen and finish labelling the system.**

Voice: A domestic hot water supply system consists of a boiler, a hot water storage tank and a cold water supply tank. These are connected to each other by pipes. Hot water leaves the boiler by flow pipe A at the top of the boiler. This pipe enters the top of the hot water tank. When hot water is taken for domestic use, it leaves the hot water tank from another pipe at the top of the hot water tank: pipe D. This water is replaced by cold water from the cold water supply tank. The water runs down pipe C, which goes from the bottom of the cold water tank to the bottom of the hot water tank. Return pipe B replaces the hot water that leaves the boiler with cooler water. It runs from the bottom of the hot water tank to the boiler. There is also an expansion pipe. This runs from the top of the hot tank and is bent twice at right angles so that its open end is above the cold tank. The expansion tank allows excess hot water steam to run into the cold water tank and lets any air that is in the system escape.

Presenter:

Lesson 9 Talking about oil and water separation

E **Listen and compare the description with yours.**

Voice: 1 The oil and water mixture flows through the inlet valve and accumulates on the left-hand side of the weir. That's when the chemicals are added.

2 The water sinks to the bottom of the tank and flows out of the water release pipe.

3 The oil floats on the surface of the water, crosses the weir and flows out of the oil release pipe.

4 The pressure of the system is maintained by venting gas through the gas relief pipe and adjusting the relief valves on the water and oil pipes.

Presenter:

Unit 9 Giving advice

Lesson 3 Comparing advice and obligation

A **A supervisor is talking to Jim, a new roustabout who is starting work on the floor of the rig. Listen to the conversation and answer the questions.**

Supervisor: Okay Jim, at the beginning it won't be very exciting. You'll have to spend a lot of time cleaning up – it's important that the surfaces are not wet or slippery. Make sure you wear the correct PPE.

Jim: Do I have to wear a hard hat and gloves all the time?

Supervisor: Well, you should wear the hard hat in this area mainly, not inside. You don't have to wear gloves all the time – but you must wear the safety boots.

Jim: It's quite a lot to remember!

Supervisor: If you're not sure, you can always ask one of the crew. It's also important to read the regulation handbook.

Jim: I had a look at it yesterday, but I can't remember everything.

Supervisor: Well, try to find time to read it again before you start work. In fact, you should read the safety regulations handbook today if possible.

Jim: I think I need a cigarette.

Supervisor: Well, don't forget, you mustn't smoke on the floor of the rig!

Presenter:

F **Listen and check your answers.**

Voice 1: He must have been stealing the tools. I'd report him to the supervisor. Yes, I think you should do that.

Voice 2: If you're tired, you should ask if you can take more breaks – just short ones. You should make sure you eat regularly and go to bed early.

Voice 3: The rules are there for a reason. You have to follow them. If you think some of them are unnecessary, perhaps you could ask your supervisor why they are important.

Voice 4: Maybe you should change your job! Or if that's not possible, be patient and see if it gets more interesting. If you show you are good at the job you have, you might be given more interesting things to do. It's a good idea to look around at other people who have the jobs you would like to do and find out what you need to do to get a job like that.

Presenter:

Lesson 4 Troubleshooting

A **Work in groups. Look at the pictures and predict what is going to happen. Listen to some people discussing each picture and see if their ideas are the same as yours.**

1

Voice 1: That plug isn't wired properly, someone will be electrocuted if it isn't fixed.

Voice 2: It's not that dangerous, I think it will give someone an electric shock, but it won't electrocute them!

2

Voice 1: Oh no, the lighted cigarette shouldn't be next to the oil drum – it's going to catch fire!

Voice 2: Yes – it may go out, but if there's a strong wind or if the oil is leaking, it might cause a serious problem.

3

Voice 1: Is that oil on the ground? It may have come from the engine.

Voice 2: Yes, it could have an oil leak. If someone doesn't fix it and top up the oil, it will destroy the engine.

 CD2 T33

Presenter:

B **Listen again and complete the sentences in the box below.**
[REPEAT OF EXERCISE A]

 CD2 T34

Presenter:

Lesson 6 Giving toolbox talks

D **Listen. What is the topic of the talk?**

Voice: I want to talk to you about using cranes, forklifts, mobile plant and other lifting machines. The first rule is that you must have the correct CTA or equivalent licence to use the machinery and you must be authorized by the site. Next, always remember to walk around the machine before you use it to check for problems or defects. Carry out proper checks on the machinery – the tyres, the lights and the brakes particularly. You also have to check for potential hazards such as overhead cables and other employees who might be in the way. Make sure you know the safe working load of your machine and the weight of the load you are going to lift. See if the machine can take the load by lifting it slightly, then halting. But you should never leave the cab while the load is suspended, and you mustn't ever stand under it. When you get in the cab, you

must wear a seat belt if it's provided – keep to the speed limit and don't allow unauthorized passengers. Never leave the cab unattended. Lastly, at the end of the day, make sure you park on level ground, and always lock the cab, windows and any covers.

 CD2 T35

Presenter:

Lesson 7 Using formal and informal technical expressions

A **John is a reporter visiting the rig where Bob works. Bob is telling him how the machinery works. Listen and finish labelling the diagram below.**

John: We're standing by the rig and I'm with Bob, who has offered to talk to us about it. Bob, can you explain the drilling process to me?

Bob: I'll have a go.

John: Now, I know this is the well with the derrick standing over it.

Bob: Yes, that's right, and you can see the rotating turntable at the top of the well with the kelly.

John: What does the kelly do?

Bob: It's a pipe that transfers the rotary movement to the rotating turntable and the drill.

John: And the drill pipe and collars are called the drill string, is that right?

Bob: Yes, that's right. The drill bit is at the end of the drill string – that's the specialized cutter that cuts up the rock.

 CD2 T36

Presenter:

Lesson 8 Speaking hypothetically

B **Listen to Bob explaining about older oil wells. Compare his answers with yours.**

Bob: When oil wells get older, it becomes more difficult to pump oil efficiently, as the pressure decreases. The recovery of oil can be improved by using water injection – it can be improved from fifteen percent to as much as fifty percent. Water injection wells are usually located at the periphery or edge of the field and by forcing water below the accumulation, the crude oil is pushed towards the producing wells near the centre. As well as raising the pressure in the wells, water injection also keeps oil below its bubble point. This is the point when gas can escape from the crude oil in great quantities and is produced more than oil. Before water is injected into a well though, it must be treated to remove solids, bacteria and oxygen that might cause damage or corrode the equipment. Sea water can be used for injection on offshore platforms, but it needs to be desalinated – to have the salt removed – because untreated sea water can be highly corrosive.

🔊 (CD2 T37)

Presenter:
C **Listen again and decide whether the statements below are true or false.**
[REPEAT OF EXERCISE B]

🔊 (CD2 T38)

Presenter:
Lesson 10 Describing filters and strainers
E **Listen and check your answers. Complete the table with the advantages and disadvantages of each type of filter.**

Voice: Cone filters are fitted within pipelines. They are quite effective and are simpler than duplex filters. The main problem with them is that they cannot be accessed easily and, if they need cleaning or repairing, it is necessary to break the pipeline. Duplex filters contain more than one filter. This means that when one filter is offline for cleaning, the flow can be directed to the other filter.

WORD LIST

ablaze	blow	chemical	convection
abrasive	blowout	chest	copper
ABS	boil	chip	core
accelerator	boiler	chisel	corrosive
accurate	boiling point	chuck	corrugated
acid	bolt	circuit	countable
add	bone	circular	counterweight
adjust	boom	circulate	crack
airbag	borehole	circulation	crane
airlock	bourdon	circumference	crew
airway	brake	clamp	crude oil
aisle	brass	clay	cubed
alarm	breathe	clockwise	cubic
alcohol	breathing	cloud	cuboid
aluminium	brittle	cloudy	current
ankle	brittleness	column	cuttings
annulus	bubble point	combustible	CV
anti-clockwise	bulb	combustion	cylindrical
approximate	buoy	commencement	debris
area	burn	competent	decade
artificial ventilation	burner	compound	decibel
asphalt	button	compress	deck
assess	buzzer	compression	demulsifier
atom	bypass	compressor	density
ATV	cable	concentric	derrick
avoid	calculate	concise	device
avoided	calipers	condense	diagnose
back-up	capillary tube	conduction	dial
bar	carbon dioxide	conductive	diameter
barge	careful	conductivity	diesel
barrier	careless	cone	discharge
bell	casing	confined	disconnect
bellow	casualty	connect	dislocation
bench	cease	construct	displacer
bit	centimetre	contamination	disposable
blade	centrifuge	contract	disposal
blend	chain saw	contusion	disturbance

diver	finish	hammer	kelly
diverted	fire extinguisher	handle	kelvin
divide	first-aid	hard hat	kerosene
doodlebugger	fish	harm	kick
downtime	fixed	harness	kilogram
drain	flame	hazard	kilometre
draw	flammable	hazardous	kilowatt
driller	flare	heart	knee
driving licence	flexible	heavy duty	kneel
drop	float	height	laceration
ductile	floorman	hertz	laminar
dull	flow	hexagonal	leak
duplex	fluid	hoist	length
durability	foam	hold	lever
durable	fog	hole	lifeline
ear protectors	foggy	horizontal	lift
elastic	fold	hose	lightning
elbow	foot	icy	lightweight
element	force	ignite	limb
elevation	forehead	inch	lip
eliminate	forklift	indicator	liquid
emergency	formation	inhale	load
emulsion	fraction	inject	locate
encourage	fracture	injure	loop
environment	fragile	injury	loosen
equals	frequently	insufficient	lower
equipment	friction	interact	LPG
exact	fuse	invar	lubrication
examine	fusibility	iron	lung
excess	fusible	isolate	lustre
expand	gasoline	isolation	lustrous
explosion	gauge	jack	maintenance record
extend	generator	jack-up	malleable
extinguish	geophones	jaw	mark
face mask	goggles	jet fuel	mass
fall	grinder	jib	mat
faulty	guard	joint	mechanism
file	gusher	jughustler	megawatt
filter	hack saw	junction	mercury
finger	hail	junk	metallurgist

meter	pigging	relay	sharpen
metre	pinch	release	shock
mile	pipeline	remove	shoulder
milligram	piston	reservoir	side-impact bars
millimeter	platform	resistant	signal
minus	pliers	responsible	slacken
mist	plug	restrain	sleet
mixture	plunger	retract	slew
modify	pointer	rib	slide
monitor	policy	rig	slope
motorman	pontoon	rigid	smooth
mud	potential	ring	socket
mudman	power	risk	soiled
multigrade lubricant	PPE	roadworthy	solid
multiple	practical	rotary hose	solute
multiply	preserve	rotary table	solvent
neck	preset	rotate	sophisticated
nozzle	press	rotor	spanner
observe	pressure	rough	spherical
offline	prevent	roughneck	spillage
offshore	primary	round trip	spiral
oilfield	procedure	roustabout	spotter
online	prohibit	rubber	sprain
onshore	promote	safety	spring
on stream	propel	sand trap	steel
operation	protect	saw	steering wheel
operator	pull	scaffolding	strain
outlet	pulse	scale	strainer
overalls	push	scenario	strike
overflow	radiation	screw	string
paraffin wax	raise	screwdriver	stroke
pascal	rarely	seal	suction
passenger	reciprocating	seat belt	sunny
pedestrian	recorder	seismic	superficial
pentane	recover	seldom	supply
percent	recovery	semi-submersible	surface
permit	rectangular	set	swing
petrol	registration document	severe	switch
petroleum	regulate	shale shaker	tag
pig	reinforce	shallow	tank

tear	viscous
tensile strength	visual display unit
tension	voltage
terminal	weir
thermal	wheel
thermostat	windy
throat	wire
thumb	worn
thunder	wrench
tighten	wrist
tilt	
toe	
toolbox (talk)	
torso	
traditional	
trainee	
tray	
treater	
treatment	
triangular	
trigger	
trip (the pipe)	
troubleshooting	
tube	
turbine	
twist	
ultrasonic	
ultrasound	
unconscious	
unidirectional	
untreated	
utilize	
valve	
vapour	
variable	
vent	
versatile	
vertical	
vibrate	
vice	